10 Virginia SOL Grade 6 Math Practice Tests

The Ultimate Test Prep Collection with Answer Explanations

Dr. A. Nazari

10 Practice Tests

👑 *The Grand Championship Collection* 👑

Welcome, future Math Champion!

*You hold the **ultimate collection** —
ten full-length practice tests designed to take you
from first attempt to **complete mastery.***

🏅 *Conquer every Grade 6 topic*

🏅 *Build unshakeable confidence*

🏅 *Rise from Bronze to Gold to Champion*

🏅 *Arrive at test day fully prepared*

The championship begins now.

★ ★ ★ ★ ★

❝ *Ten tests may seem like
a marathon, but champi-
ons are made one step at a
time. Trust the process!* **❞**

The Champion's Path

Your 4-phase journey from Bronze to Champion

Bronze Round (Tests 1–3)

Your warm-up matches. Take these **untimed** to learn the format and set your baseline. Read the answer explanations after each test — this is where you build your foundation.

Silver Round (Tests 4–6)

Set a timer for **75 minutes**. Focus on the topics that tripped you up in Bronze. Practice showing your work on every problem. Your accuracy should be climbing.

Gold Round (Tests 7–9)

Full timed conditions (**60 minutes**). Simulate the real exam environment. Review only the questions you missed — targeted practice is the key to gold.

Championship Final (Test 10)

Your final match. Full exam conditions — timed, quiet, no breaks. This is your victory lap. Show yourself how far you've come!

Your Championship Kit

- **10 Full-Length Practice Tests** — every Grade 6 topic
- **Formula Reference Sheet**
- **Complete Answer Key** with explanations
- **Championship Scoreboard** to track your rise

Champion's Tip: Space your tests 2–3 days apart. Use the days in between for targeted review. By Test 10, you'll be amazed at your transformation.

👑 The Champion's Playbook 👑

I   **Read every question twice.** The first read tells you the topic. The second tells you exactly what to solve for. Champions never skim.

II   **Mark the clues.** Circle key numbers, underline the question, and cross out information that's just there to distract you.

III   **Choose your strategy.** Before touching pencil to paper, decide: Am I setting up a ratio? Solving an equation? Finding area? Name the approach.

IV   **Solve, then match.** For multiple choice — work the problem on scratch paper first, then find your answer among the choices.

V   **Eliminate and conquer.** Cross out obviously wrong answers. If you're left with two, you've already doubled your odds. Make an educated pick.

VI   **Estimate to verify.** After solving, ask: "Is this answer reasonable?" A quick mental estimate catches most calculation errors.

VII   **Leave nothing blank.** Even a well-reasoned guess is worth more than an empty space. Use partial work to support your answer.

Tests 1–3: **Untimed** (build foundation)   Tests 4–6: **75 min** (build speed)   Tests 7–10: **60 min** (championship conditions)

⭐ *Grade 6 Championship Topics*

🥇 Ratios & Proportions   🥇 Integers & Rational Numbers   🥇 Expressions & Equations

🥇 Geometry & Measurement   🥇 Statistics & Data Analysis

*A true champion isn't someone who never makes mistakes — it's someone who learns from **every single one**. After each test, review your errors carefully. That's where the real growth happens.*

Find more at
ViewMath.com/VA-Grade6

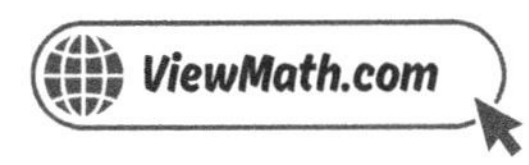

♔ The Champion's Toolkit ♔

Prepare your workspace before each championship round

🏆 Required Equipment

Sharpened Pencils — Two #2 pencils — champions always have a backup

Quality Eraser — A clean, soft eraser that won't smudge your work

Scratch Paper — Blank paper for calculations, diagrams, and number lines

Ruler — Essential for geometry and coordinate plane questions

Timer — Begin using from the Silver Round onward

Quiet Workspace — A calm, well-lit area free from distractions

Allowed in Competition	Not Allowed
✓ Pencil and eraser	✗ Calculators
✓ Scratch paper (provided)	✗ Electronic devices
✓ Ruler (if specified)	✗ Textbooks or notes
✓ Formula reference in this book	✗ Outside help

👥 *For Parents & Teachers*

- *With 10 tests, space them **2–3 days apart**. This gives time to review mistakes and study between rounds.*

- *Let your child take Tests 1–3 untimed to build familiarity and establish a baseline.*

- *After each test, go through the Answer Key together. Focus on **understanding the reasoning**, not memorizing answers.*

- *Use the Championship Scoreboard to visualize long-term progress. Celebrate improvements at every tier!*

- *Pair with our **Grade 6 Math Study Guide** for topics that need sustained attention.*

Formula Reference Sheet

◢ Area Formulas

Rectangle	$A = l \times w$
Parallelogram	$A = b \times h$
Triangle	$A = \dfrac{1}{2} \times b \times h$
Trapezoid	$A = \dfrac{1}{2}(b_1 + b_2) \times h$

▣ Volume

Rectangular $V = l \times w \times h$

Prism

▣ Surface Area

Find the area of each face, then add them all up.

Rectangular Prism:

$SA = 2lw + 2lh + 2wh$

↓ Order of Operations

P	Parentheses first
E	Exponents
M/D	Multiply & Divide (left to right)
A/S	Add & Subtract (left to right)

％ Ratios & Percents

Ratio: $a : b$ or $\dfrac{a}{b}$

Unit rate: amount per 1 unit

Percent: a ratio out of 100

Part = Percent $\times$ Whole

⚖ Integers & Absolute Value

Integers:

$$\ldots, -3, -2, -1, 0, 1, 2, 3, \ldots$$

$$|-5| = 5 \quad |5| = 5$$

Absolute value = distance from 0

X¹ Expressions & Equations

Exponent: $3^4 = 3 \times 3 \times 3 \times 3 = 81$

Variable: a letter that stands for a number

Equation: two expressions joined by $=$

Inequality: uses $<, >, \leq, \geq$

⊕ Coordinate Plane

Ordered pair: (x, y)

x-axis: horizontal y-axis: vertical

Origin: $(0, 0)$

Four quadrants (I, II, III, IV)

Statistics

Mean: sum of values $\div$ count

Median: middle value (sorted)

Range: max $-$ min

 Championship Scoreboard

Track your rise through every championship round

Champion's Name: _______________________________

🏆 Round	🏅 Tier	📅 Date	⭐ Score	😊 Rating
1	Bronze			
2	Bronze			
3	Bronze			
4	Silver			
5	Silver			
6	Silver			
7	Gold			
8	Gold			
9	Gold			
10	👑			

X

My strongest topics (where I consistently score well):

____

____

Topics I improved on the most from Bronze to Gold:

____

____

My score trend (Bronze avg → Gold avg → Championship):

____

____

One strategy that helped me improve the most:

____

____

My confidence level for the real test (1–10):   _________ / 10

Continue Learning at
ViewMath Academy!

For Parents, Teachers & Students

Great job on the practice tests! Want to keep improving? ViewMath Academy is your **free online companion** to this book.

- **Score Analyzer** — Enter your answers and instantly see which topics need more practice
- **Interactive Lessons** — Review the concepts behind each question with clear explanations
- **Adaptive Quizzes** — Practice your weak topics with questions that match your level
- **Progress Tracking** — See your mastery grow across all Grade 6 math topics
- **Personalized Dashboard** — A learning plan tailored just for you

Scan to visit ViewMath Academy

ViewMath.com/VA-Grade6

 Free to use · No downloads required · Works on any device 

⭐ Table of Contents ⭐

Here's what we'll explore together!

Let's learn and have fun!

Practice Test 1

30 Questions

✏ Before You Start ✏

- ✓ **Read each question carefully** before choosing your answer.
- ✓ **Show your work** on scratch paper when you need to.
- ✓ **Skip hard questions** and come back to them later.
- ✓ **Check your answers** when you're done.
- ✓ **Take your time** — there's no rush!

 You've Got This! 

Do your best and show what you know!

1. A garden has flowers and vegetables in a ratio of 4 : 1. There are 20 flowers. How many vegetables are there?

2. The graph shows the relationship between hours worked and money earned.

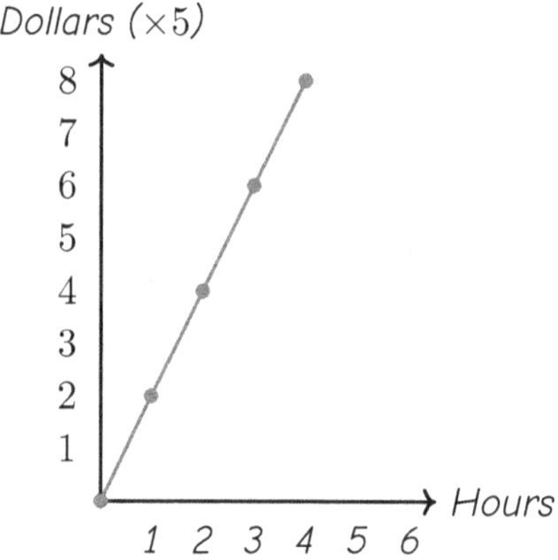

How much money is earned in 6 hours?

A) $30

B) $50

C) $60

D) $40

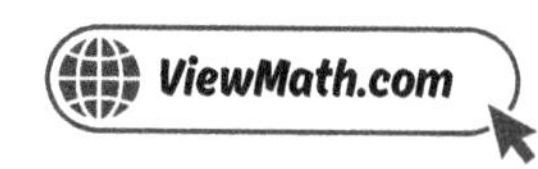

3. *The table and partial graph below show a ratio relationship.*

x	y
1	4
2	8
3	?

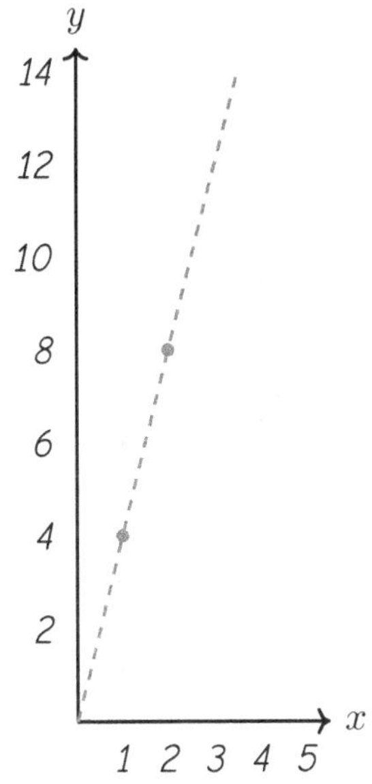

What is the missing value of y when $x = 3$?

(A) 10

(B) 12

(C) 14

(D) 16

4. *What is 15% of 60?*

(A) 9

(B) 15

(C) 45

(D) 6

5. *A student solved the problem below using "Keep, Change, Flip." Look at the student's work.*

Problem: $\dfrac{4}{5} \div \dfrac{2}{3}$

Step 1: Keep $\dfrac{4}{5}$ Change $\div$ to $\times$

Step 2: Flip $\dfrac{4}{5}$ to get $\dfrac{5}{4}$

Step 3: $\dfrac{5}{4} \times \dfrac{2}{3} = \dfrac{10}{12} = \dfrac{5}{6}$

What error did the student make?

(A) The student forgot to change $\div$ to $\times$

(B) The student made a multiplication error in Step 3

(C) The student forgot to simplify

(D) The student flipped the wrong fraction

6. *Compute* $0.288 \div 0.12$.

Your Answer

7. *The number line below marks the multiples of 3 with circles (●) and the multiples of 4 with triangles (▲).*

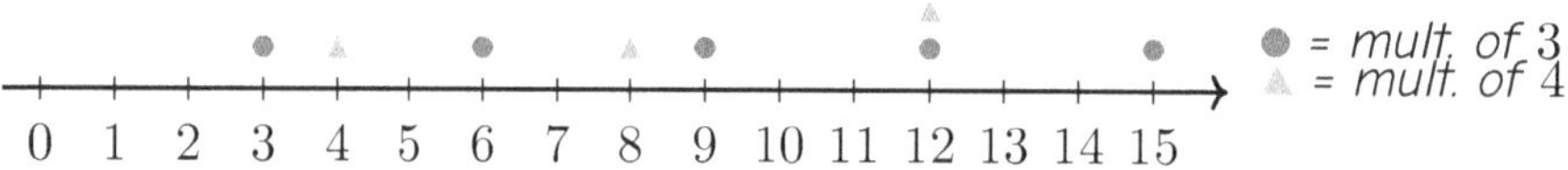

What is the LCM of 3 and 4?

(A) 1

(B) 4

(C) 12

(D) 24

8. *Use the distributive property to find* 8×27.

(A) 216

(B) 206

(C) 196

(D) 226

Find more at
ViewMath.com/VA-Grade6

ViewMath.com

9. A point is at $(6, -2)$. It is reflected across the y-axis, then reflected across the x-axis. What are the coordinates of the final image?

Your Answer:

10. On a town map, the park is at $(-6, 4)$ and the museum is at $(2, 4)$. Each grid unit represents one block. How many blocks apart are the park and the museum? Show your work.

Your Answer:

11. The diagram below shows an expression broken into its parts. How many terms does the expression have?

$$\boxed{7m} \; + \; \boxed{2n} \; - \; \boxed{4} \; + \; \boxed{3m}$$

(A) 2

(B) 3

(C) 4

(D) 5

12. The triangle below has a base of $b = 10$ cm and a height of $h = 6$ cm. Use the formula $A = \dfrac{1}{2} \times b \times h$ to find its area.

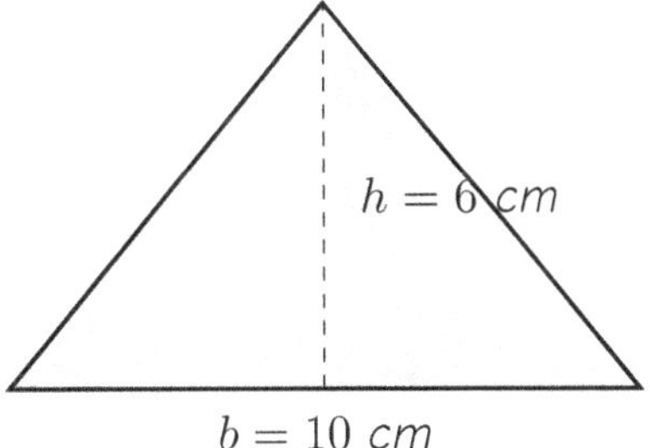

Your Answer:

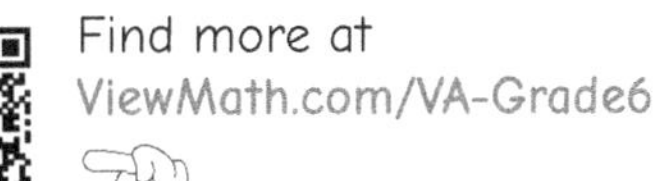

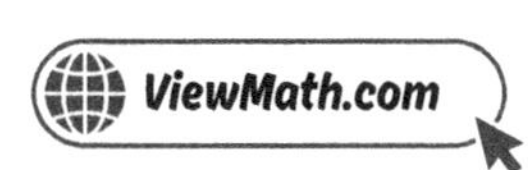

13. *Factor using the distributive property:* $15n + 10$

14. *A gym charges \$25 to join and \$10 per month. Which expression gives the total cost after m months?*

   (A) $25m + 10$

   (B) $25 + 10m$

   (C) $35m$

   (D) $25 - 10m$

15. *Name three solutions to the inequality $m \geq 6$.*

16. *Which inequality matches a graph with an open circle at -2 and shading to the left?*

   (A) $x > -2$

   (B) $x < -2$

   (C) $x \geq -2$

   (D) $x \leq -2$

17. *A triangular flower bed has a base of 4.5 m and a height of 6 m. What is its area?*

18. *A parallelogram has an area of 96 ft^2 and a height of 8 ft. What is the base?*

Find more at
ViewMath.com/VA-Grade6

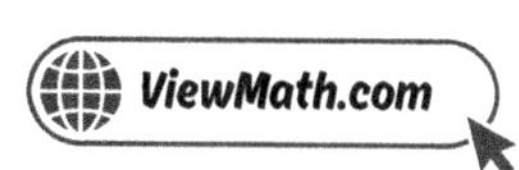

19. Two boxes have the same volume. Box A is $6 \times 4 \times 5$. Box B is $10 \times 3 \times h$. What is the height h of Box B?

(A) 2

(B) 4

(C) 6

(D) 8

20. If a triangle is reflected across the x-axis, what stays the same?

(A) The y-coordinates of all vertices

(B) The size and shape of the triangle

(C) The direction the triangle faces

(D) The sign of the y-coordinates

21. A right triangle has vertices $(-4, 0)$, $(2, 0)$, and $(-4, 5)$. What is the area?

(A) 30 square units

(B) 15 square units

(C) 11 square units

(D) 20 square units

22. A gift box is 12 in long, 8 in wide, and 5 in tall. How much wrapping paper is needed to cover all sides?

(A) 392 in^2

(B) 480 in^2

(C) 296 in^2

(D) 196 in^2

23. A point at $(2, -6)$ is reflected across the y-axis. Where does it land?

(A) $(2, 6)$

(B) $(-2, 6)$

(C) $(-2, -6)$

(D) $(6, -2)$

Find more at
ViewMath.com/VA-Grade6

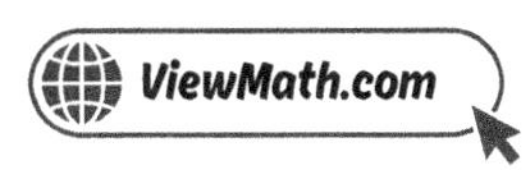

24. *A circular pool has a diameter of 20 feet. What is the area of a pool cover that fits exactly over the pool? Use* $\pi \approx 3.14$

(A) 62.8 ft^2 (B) 157 ft^2

(C) 314 ft^2 (D) 1,256 ft^2

25. *Data:* 10, 20, 30, 40, 50, 60. *What is the median?*

(A) 30   (B) 33

(C) 35   (D) 40

26. *The table below shows the distances (in miles) five students travel to school and their distances from the mean.*

Student	Miles	Distance from Mean
A	2	4
B	4	2
C	6	0
D	8	2
E	10	4

What is the MAD?

(A) 0   (B) 2

(C) 2.4   (D) 6

27. *A histogram has bars for 0–9, 10–19, 20–29, 30–39 with heights* 3, 10, 8, 4. *Which interval contains the most data?*

(A) 0–9   (B) 10–19

(C) 20–29   (D) 30–39

Find more at
ViewMath.com/VA-Grade6

28. A student says "The data set with the larger range is always more spread out than the one with the smaller range." Is this always true? Explain.

29. A standard number cube is rolled. What is the probability of rolling a number less than or equal to 3?

(A) $\dfrac{1}{3}$

(B) $\dfrac{3}{6}$

(C) $\dfrac{2}{6}$

(D) $\dfrac{4}{6}$

30. A family earns $3,000 per month. Their circle graph shows 15% goes to entertainment. How many dollars go to entertainment each month?

Find more at
ViewMath.com/VA-Grade6

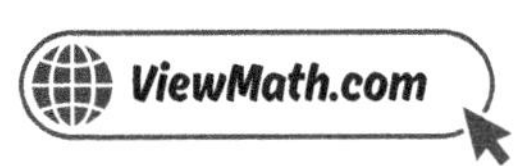

⭐ End of Practice Test 1 ⭐

Great job finishing the test!

 **My Score**

I got ___________ out of 30 questions right.

Check your answers in the Answer Key at the back of the book.

 Review any questions you missed. That's how we learn!

📊 Check Your Score Online!

Visit **ViewMath Academy** to enter your answers and see which topics you need to review. You can also explore lessons, take quizzes, track your scores, and save your progress!

viewmath.com/score/6.1.VA.16

Or go to viewmath.com/score and enter code: 6.1.VA.16

2

Practice Test 2

☑ 30 Questions

✏ **Before You Start** ✏

- ✓ **Read each question carefully** before choosing your answer.
- ✓ **Show your work** on scratch paper when you need to.
- ✓ **Skip hard questions** and come back to them later.
- ✓ **Check your answers** when you're done.
- ✓ **Take your time** — there's no rush!

⭐ You've Got This! ⭐

Do your best and show what you know!

1. Look at the tape diagram below.

Boys: ☐ ☐ ☐
Girls: ☐ ☐ ☐ ☐ ☐ ☐

There are 24 girls. How many boys are there?

(A) 4

(B) 12

(C) 16

(D) 20

2. Complete the ratio table below. The ratio of markers to crayons is $4 : 6$.

Markers	Crayons
4	6
8	?
?	24
20	?

Find all three missing values.

3. Two runners' distances are graphed below.

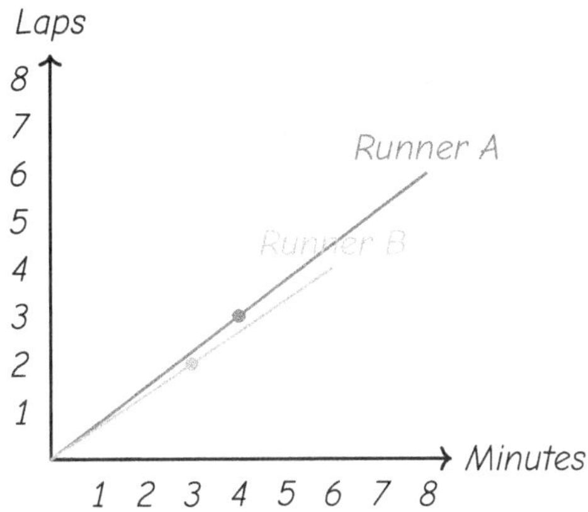

Part A: What is each runner's ratio of laps to minutes?

Part B: Who runs faster? Explain using the graph.

Your Answer

4. 25% of what number is 20?

(A) 5

(B) 40

(C) 80

(D) 100

5. A student solved $\dfrac{3}{5} \div \dfrac{2}{9}$ and got $\dfrac{6}{45}$. What mistake did the student make?

(A) The student forgot to simplify the answer

(B) The student multiplied without using the reciprocal

(C) The student flipped the first fraction instead of the second

(D) The student added the fractions

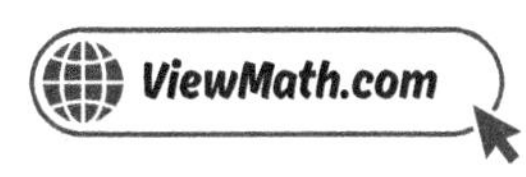

6. A store sells ribbon at $0.60 per meter. How much does 3.5 meters of ribbon cost?

   (A) $2.10

   (B) $2.16

   (C) $21.00

   (D) $0.21

7. What is the LCM of 4 and 6?

   (A) 2

   (B) 12

   (C) 24

   (D) 48

8. Marcus wrote $54 + 36 = 6(9 + 4)$. Explain the error and write the correct factored form using the GCF.

9. Starting at the origin, how do you plot the point $(-3, 4)$?

   (A) Move 3 units right and 4 units up

   (B) Move 3 units left and 4 units down

   (C) Move 3 units left and 4 units up

   (D) Move 4 units left and 3 units up

10. The points $(5, -1)$ and $(-3, -1)$ lie on a horizontal line. Explain step by step how to find the distance between them, and state why the answer is positive.

11. Which part of the expression $7x^2 + 3x - 5$ is a constant?

   (A) 7

   (B) x^2

   (C) $3x$

   (D) 5

Find more at
ViewMath.com/VA-Grade6

12. Evaluate $m^2 + 2m + 1$ when $m = 4$.

Your Answer:

13. The rectangle below has the dimensions shown. Write a simplified expression for its perimeter.

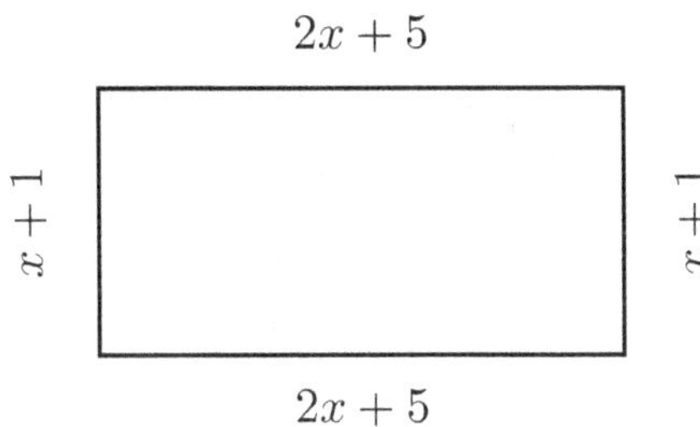

14. Write a situation that could be modeled by the expression $5n + 3$.

15. You must be at least 48 inches tall to ride a roller coaster. Which inequality represents the required height h?

A) $h > 48$

B) $h < 48$

C) $h \leq 48$

D) $h \geq 48$

16. What is the difference between the graphs of $x > 5$ and $x \geq 5$?

A) They shade in different directions

B) $x > 5$ uses an open circle; $x \geq 5$ uses a closed circle

C) $x > 5$ shades right; $x \geq 5$ shades left

D) There is no difference

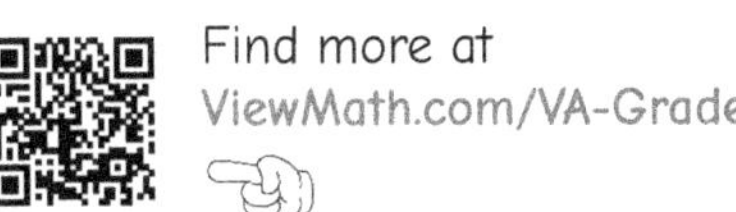

17. *A triangular garden has a base of 12 ft and a height of 9 ft. How many square feet of soil are needed to cover it?*

(A) 108 ft²

(B) 54 ft²

(C) 21 ft²

(D) 42 ft²

18. *A parallelogram has base 6.5 ft and height 4 ft. What is the area?*

(A) 26 ft²

(B) 13 ft²

(C) 21 ft²

(D) 10.5 ft²

19. *A cube has edge length 6 in. What is its volume?*

20. *What is the reflection of $(4, 7)$ across the x-axis?*

(A) $(-4, 7)$

(B) $(4, -7)$

(C) $(-4, -7)$

(D) $(7, 4)$

21. *A rectangle on the coordinate plane has area 54 square units. Two of its vertices are at $(-3, 2)$ and $(6, 2)$. What is the width of the rectangle?*

(A) 6 units

(B) 9 units

(C) 3 units

(D) 18 units

22. *A classroom has dimensions 10 m by 8 m by 3 m. What is the total surface area of the walls, floor, and ceiling?*

23. *A point is translated 4 units right and 3 units down. Which rule describes this translation?*

(A) $(x, y) \to (x + 4, y + 3)$

(B) $(x, y) \to (x - 4, y - 3)$

(C) $(x, y) \to (x + 4, y - 3)$

(D) $(x, y) \to (x - 4, y + 3)$

24. *A circular tablecloth has a diameter of 6 feet. Lace trim costs \$2 per foot. How much will it cost to add lace trim around the entire edge of the tablecloth? Use $\pi \approx 3.14$.*

Your Answer:

25. *The bar graph below shows test scores for 5 students.*

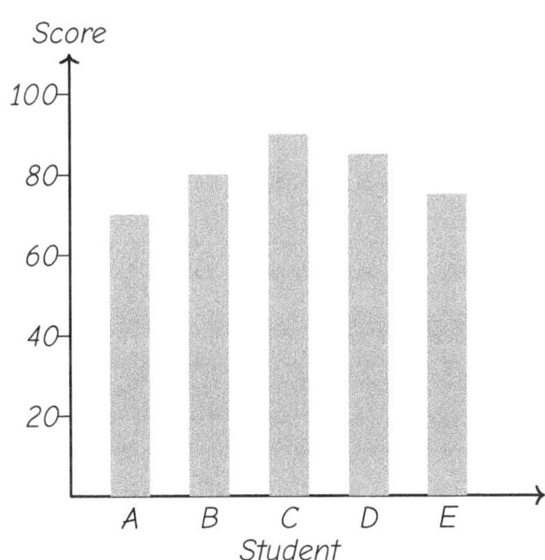

The scores are 70, 80, 90, 85, 75. What is the mean score?

(A) 75

(B) 80

(C) 85

(D) 90

26. The dot plot below shows the number of laps students ran during gym class.

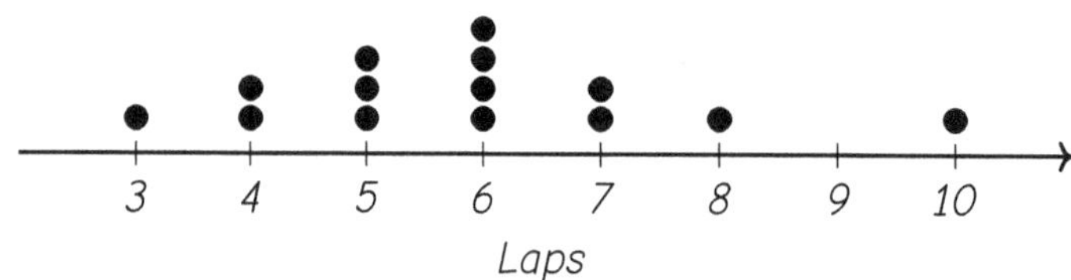

Find the range and IQR from the dot plot data.

27. A frequency table lists values and how many times each appears. Another name for the value that appears most often is —

(A) mean

(B) median

(C) mode

(D) range

28. A coach records sprint times (seconds) for two athletes over 10 races. Athlete A: median $= 12.5$, range $= 1.2$. Athlete B: median $= 12.5$, range $= 3.0$. Which athlete is more consistent?

(A) Athlete A

(B) Athlete B

(C) They are equally consistent.

(D) Cannot be determined without the IQR.

29. The probability scale below shows four events. Which event has a probability closest to 0.75?

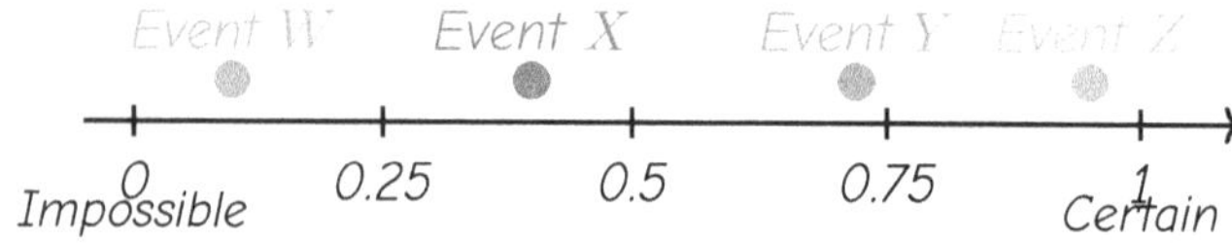

(A) Event W

(B) Event X

(C) Event Y

(D) Event Z

30. *The circle graph below shows how a family spends its monthly income.*

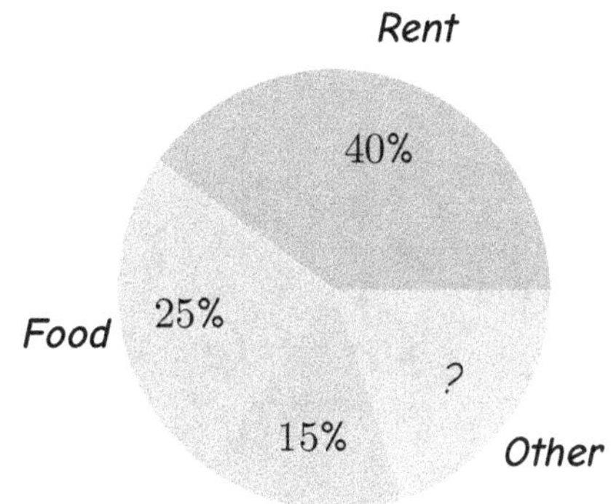

*What percent of the family's income goes to the **Other** category?*

(A) 10%   (B) 15%

(C) 20%   (D) 25%

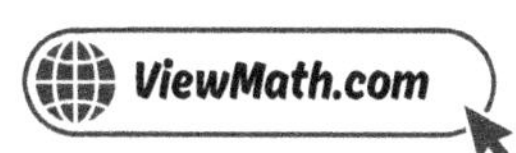

⭐ *End of Practice Test 2* ⭐

Great job finishing the test!

 *My Score*

I got _____________ out of 30 questions right.

Check your answers in the Answer Key at the back of the book.

Review any questions you missed. That's how we learn!

📊 Check Your Score Online!

Visit **ViewMath Academy** to enter your answers and see which topics you need to review. You can also explore lessons, take quizzes, track your scores, and save your progress!

viewmath.com/score/6.1.VA.17

Or go to *viewmath.com/score* and enter code *6.1.VA.17*

3

Practice Test 3

30 Questions

✏️ Before You Start ✏️

- ✓ **Read each question carefully** before choosing your answer.
- ✓ **Show your work** on scratch paper when you need to.
- ✓ **Skip hard questions** and come back to them later.
- ✓ **Check your answers** when you're done.
- ✓ **Take your time** — there's no rush!

⭐ You've Got This! ⭐

Do your best and show what you know!

1. A baker uses 2 eggs **per** 3 cups of flour. Which ratio represents flour to eggs?

   (A) $2:3$

   (B) $3:5$

   (C) $3:2$

   (D) $2:5$

2. A recipe uses 4 eggs for every 6 cups of flour. How many eggs are needed for 18 cups of flour?

   (A) 6

   (B) 10

   (C) 12

   (D) 8

3. A ratio graph passes through $(6, 2)$. What is the value of y when $x = 15$?

   (A) 3

   (B) 5

   (C) 7

   (D) 10

4. A store sells a book for \$25. Next week the price goes up by 10%. What is the new price?

   (A) \$35

   (B) \$27.50

   (C) \$26

   (D) \$22.50

5. Each serving of juice is $\frac{2}{5}$ cup. A jug holds $\frac{4}{5}$ cup. How many servings are in the jug?

   (A) $\frac{8}{25}$

   (B) $\frac{2}{5}$

   (C) $\frac{5}{2}$

   (D) 2

6. *What is* $20.3 - 8.57$*?*

   (A) 12.27   (B) 11.53

   (C) 11.73   (D) 28.87

7. *Which of the following is* **not** *a common factor of* 24 *and* 36*?*

   (A) 4   (B) 8

   (C) 6   (D) 12

8. *Which expression shows* $24 + 18$ *rewritten using the GCF and the distributive property?*

   (A) $2(12 + 9)$ (B) $3(8 + 6)$

   (C) $6(4 + 3)$ (D) $6(4 + 6)$

9. *Where is the point* $(-6, 0)$ *located on the coordinate plane?*

   (A) *On the* y*-axis*   (B) *On the* x*-axis*

   (C) *In Quadrant II*   (D) *In Quadrant III*

10. *Which pair of points has the greatest distance between them?*

    (A) $(1, 3)$ *and* $(1, 7)$ (B) $(-2, 5)$ *and* $(4, 5)$

    (C) $(0, -3)$ *and* $(0, 2)$ (D) $(-1, 4)$ *and* $(3, 4)$

11. *Which expression has exactly 4 terms?*

    (A) $2x + 3y$ (B) $a + b + c$

    (C) $5m - 2n + 7 + m$ (D) $4p$

12. Evaluate $\dfrac{x^2}{4}$ when $x = 8$.

Your Answer

13. Are $4(n+2)$ and $4n+8$ equivalent?

(A) Yes — they always give the same value

(B) No — they only match when $n = 2$

(C) No — $4(n+2) = 4n+2$

(D) Yes — but only for positive values of n

14. A dog walker earns \$7 per dog. Write an expression for the earnings from walking d dogs, and find the earnings from walking 8 dogs.

15. Which of the following values is NOT a solution to $y \geq 3$?

(A) $y = 3$

(B) $y = 4$

(C) $y = 100$

(D) $y = 2.5$

16. Which of the following graphs represents $x \leq -1$?

(A) Open circle at -1, shade right

(B) Closed circle at -1, shade right

(C) Open circle at -1, shade left

(D) Closed circle at -1, shade left

17. The triangle below is drawn inside a 6×4 rectangle. What is the area of the triangle?

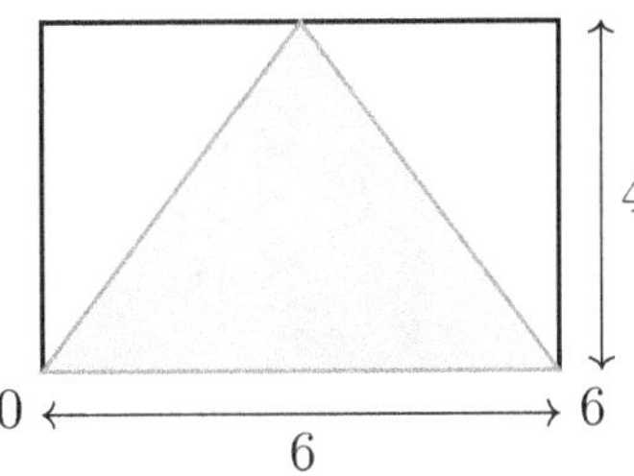

- A) 24 square units
- B) 10 square units
- C) 12 square units
- D) 18 square units

18. In a parallelogram, which measurement is used as the height?

- A) The slanted side
- B) The diagonal
- C) The perpendicular distance between the base and its opposite side
- D) The longest side

19. A storage container is 5 m long, 3 m wide, and 2 m tall. It is half-full of sand. How many cubic meters of sand are in the container?

- A) $30 \ m^3$
- B) $15 \ m^3$
- C) $10 \ m^3$
- D) $60 \ m^3$

20. A triangle has vertices $A(1,3)$, $B(5,3)$, and $C(5,7)$. What are the vertices after reflecting across the x-axis?

- A) $A'(1,-3)$, $B'(5,-3)$, $C'(5,-7)$
- B) $A'(-1,3)$, $B'(-5,3)$, $C'(-5,7)$
- C) $A'(-1,-3)$, $B'(-5,-3)$, $C'(-5,-7)$
- D) $A'(1,3)$, $B'(5,3)$, $C'(5,7)$

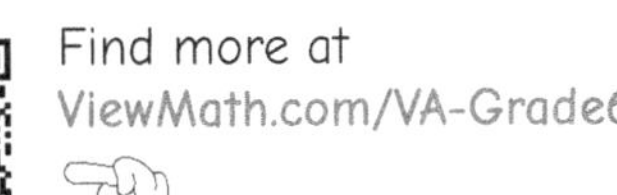

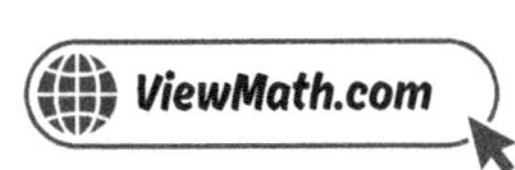

21. *A rectangle has vertices $(0,0)$, $(7,0)$, $(7,3)$, and $(0,3)$. A right triangle is cut from one corner with legs of 2 and 3. What is the area of the remaining shape?*

(A) 18 *square units*

(B) 21 *square units*

(C) 24 *square units*

(D) 15 *square units*

22. *The net below shows a triangular prism. The triangles are right triangles with legs 3 cm and 4 cm. The three rectangles have widths matching the triangle sides and a shared length of 8 cm. Find the total surface area.*

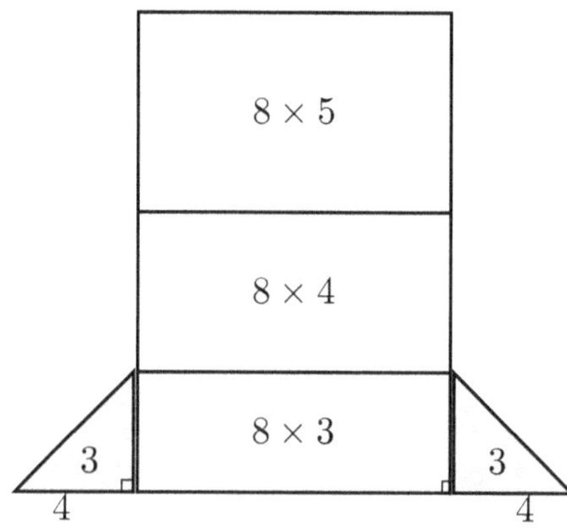

23. *After a translation, which of the following is always true?*

(A) *The shape gets larger.*

(B) *The shape flips to a mirror image.*

(C) *The shape changes its angles.*

(D) *The shape keeps the same size, shape, and orientation.*

24. *Which formula gives the area of a circle?*

(A) $A = 2\pi r$

(B) $A = \pi d$

(C) $A = \pi r^2$

(D) $A = 2\pi r^2$

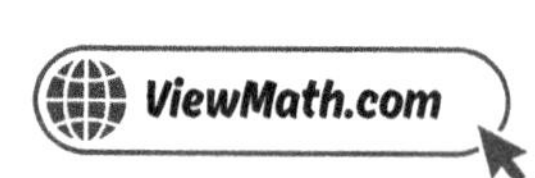

25. *The mean of 6 quiz scores is 15. If one score of 9 is removed, what is the mean of the remaining 5 scores?*

(A) 14.2

(B) 15

(C) 16.2

(D) 18

26. *Data:* 50, 55, 60, 65, 70. *The mean is 60. Find the MAD.*

27. *A data set is:* 5, 5, 5, 5, 5. *What is the mode?*

(A) 0

(B) 1

(C) 5

(D) *There is no mode.*

28. *A data set has a range of 40 but an IQR of only 8. What does this tell you?*

(A) *All values are close together.*

(B) *Most values are clustered, but there are extreme values far from the center.*

(C) *The median equals the mean.*

(D) *The data is symmetric.*

29. *A standard number cube (die) is rolled once. What is the probability of rolling a 4?*

(A) $\dfrac{4}{6}$

(B) $\dfrac{1}{4}$

(C) $\dfrac{1}{6}$

(D) $\dfrac{1}{3}$

30. *Emma earns \$40 per month in allowance. Her circle graph shows how she uses it: Savings 30%, Snacks 20%, Books 15%, Charity 10%, Fun 25%. How much does Emma put toward Savings and Charity* **combined***?*

(A) \$10

(B) \$12

(C) \$16

(D) \$20

Find more at
ViewMath.com/VA-Grade6

 # End of Practice Test 3 

Great job finishing the test!

 ## My Score

I got ___________ out of 30 questions right.

Check your answers in the Answer Key at the back of the book.

Review any questions you missed. That's how we learn!

Check Your Score Online!

Visit **ViewMath Academy** to enter your answers and see which topics you need to review. You can also explore lessons, take quizzes, track your scores, and save your progress!

viewmath.com/score/6.1.VA.18

Or go to viewmath.com/score and enter code: 6.1.VA.18

Practice Test 4

 30 Questions

---

✏️ Before You Start ✏️

- ✓ **Read each question carefully** before choosing your answer.
- ✓ **Show your work** on scratch paper when you need to.
- ✓ **Skip hard questions** and come back to them later.
- ✓ **Check your answers** when you're done.
- ✓ **Take your time** — there's no rush!

 You've Got This! 

Do your best and show what you know!

1. A store sells 3 pencils **for every** 1 eraser. Which ratio represents pencils to erasers?

   (A) $1:3$

   (B) $3:1$

   (C) $3:4$

   (D) $1:4$

2. Ella mixes paint in a ratio of 3 parts red to 5 parts white. She uses 24 parts of red. How many total parts of paint does she have?

3. A line passes through the origin and the point $(5, 15)$. What is the ratio x to y?

   (A) $1:5$

   (B) $5:1$

   (C) $1:3$

   (D) $3:1$

4. A school has 500 students. 72% passed the state exam. How many students passed?

5. What is $\dfrac{1}{2} \div \dfrac{1}{4}$?

   (A) $\dfrac{1}{8}$

   (B) 8

   (C) 2

   (D) $\dfrac{1}{2}$

6. What is 1.2×0.3?

   (A) 3.6

   (B) 0.36

   (C) 36

   (D) 0.036

Get Online

Find more at
ViewMath.com/VA-Grade6

7. You have a ribbon that is 72 inches long and another that is 96 inches long. You want to cut both ribbons into equal-length pieces with no ribbon left over. What is the longest possible piece?

(A) 24

(B) 8

(C) 288

(D) 12

8. Use the distributive property to calculate 9×34. Show your work.

9. Which of the following points is in Quadrant III?

(A) $(-1, 5)$

(B) $(3, -2)$

(C) $(-4, -7)$

(D) $(6, 1)$

10. What is the distance between $(2, -3)$ and $(2, 5)$?

(A) -8

(B) 2

(C) 4

(D) 8

11. In the expression $8(x + y)$, how many factors does the expression have?

(A) 1

(B) 2

(C) 3

(D) 4

12. The area of a triangle is $A = \dfrac{1}{2} \times b \times h$. Find A when $b = 12$ and $h = 5$.

13. *The table below shows values for two expressions. Based on the table, is Expression A equivalent to Expression B?*

x	**Expression A**: $2(x+3)$	**Expression B**: $2x+6$
0	6	6
1	8	8
5	16	16
10	26	26

- (A) *No — they only agree for small values of x*
- (B) *Yes — the distributive property shows $2(x+3) = 2x+6$ for all x*
- (C) *No — the table shows different values*
- (D) *Cannot be determined from the table*

14. *Marcus has n baseball cards. He buys 12 more and then gives 5 to his friend. Write an expression for how many cards he has now.*

15. *Which inequality represents "a number y is fewer than 20"?*

- (A) $y > 20$
- (B) $y \geq 20$
- (C) $y < 20$
- (D) $y \leq 20$

16. *A student graphed $x \geq 3$ as shown below. Is the graph correct?*

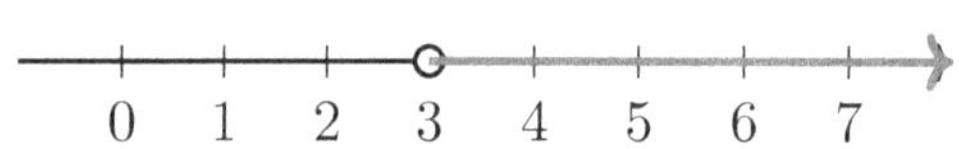

- (A) *Yes — the graph is correct*
- (B) *No — should shade to the left*
- (C) *No — should use a closed circle at 3*
- (D) *No — should use an open circle at a different number*

17. *Two triangles both have a base of 10 cm. Triangle P has a height of 6 cm and Triangle Q has a height of 8 cm. How much greater is the area of Triangle Q?*

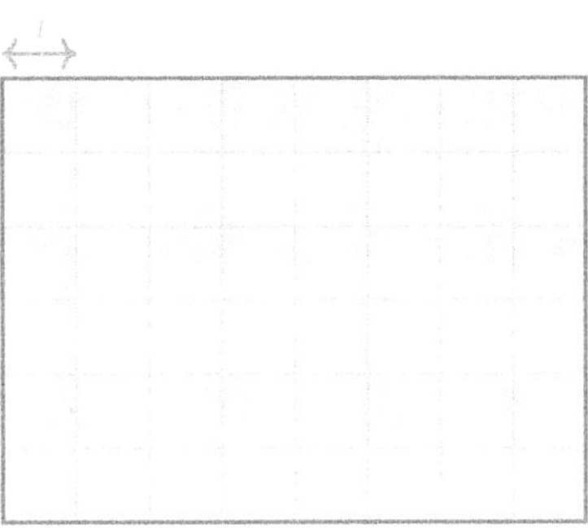

(A) 2 cm²   (B) 10 cm²

(C) 20 cm²   (D) 40 cm²

18. *A parallelogram has base 11 cm and height 3 cm. Maria says the area is 16.5 cm². What did Maria do wrong?*

(A) She divided by 2 instead of multiplying.   (B) She added the base and height.

(C) She multiplied correctly; 16.5 cm² is correct.   (D) She used the wrong unit.

19. *If you double the height of a rectangular prism but keep the length and width the same, what happens to the volume?*

(A) The volume stays the same.   (B) The volume doubles.

(C) The volume triples.   (D) The volume quadruples.

20. *After reflecting $(2, 5)$ across the x-axis, how far is the reflected point from the original?*

(A) 2 units   (B) 5 units

(C) 7 units   (D) 10 units

21. A right triangle has vertices $(2, 1)$, $(2, 7)$, and $(8, 1)$. What is the area?

   Ⓐ 36 *square units*   Ⓑ 18 *square units*

   Ⓒ 12 *square units*   Ⓓ 24 *square units*

22. A triangular prism has 2 triangular bases and 3 rectangular faces. How many faces does it have in total?

   Ⓐ 3   Ⓑ 4

   Ⓒ 5   Ⓓ 6

23. Triangle ABC is shown below. It is translated 4 units right and 2 units down. What are the coordinates of C'?

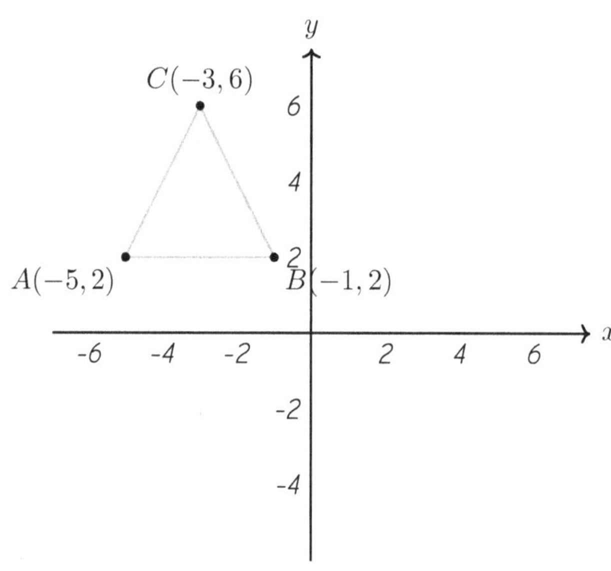

   Ⓐ $(-3, 4)$ Ⓑ $(1, 4)$

   Ⓒ $(1, 8)$ Ⓓ $(-7, 4)$

24. A sprinkler waters a circular area with a radius of 8 meters. What is the total area that the sprinkler covers? Use $\pi \approx 3.14$.

Your Answer

25. Find the median of $12, 18, 20, 24$.

  (A) 18   (B) 19

  (C) 20   (D) 21

26. Data A: $IQR = 4$. Data B: $IQR = 20$. What can you conclude?

  (A) Data A has a higher median.   (B) Data B has a higher median.

  (C) The middle 50% of Data A is more tightly clustered than Data B.   (D) Data A has more data points.

27. You want to know whether to display data as a dot plot, histogram, or frequency table. Which question should you ask first?

  (A) What is the mean of the data?   (B) How many data points are there, and is the data numerical or categorical?

  (C) What color should the bars be?   (D) Is the data symmetric or skewed?

28. Data: $12, 15, 18, 20, 22, 25, 28, 30$. What is the IQR?

  (A) 10   (B) 11

  (C) 18   (D) 8

29. *Use the probability scale below. Place each event at the correct position by writing its letter on the scale. Then state each probability as a fraction or decimal.*

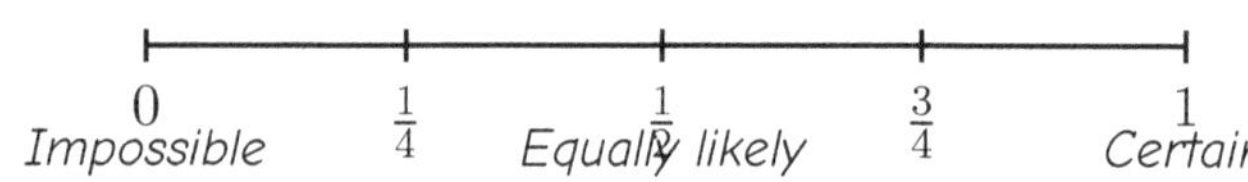

Event A: *Rolling a 7 on a standard number cube.*

Event B: *Flipping tails on a fair coin.*

Event C: *Drawing a red marble from a bag that has 6 red and 2 blue marbles.*

Event D: *Rolling a number less than 7 on a standard number cube.*

30. *A circle graph of an election shows: Candidate A received 28%, Candidate B received 34%, and Candidate C received the rest. What percent of the votes went to Candidate C?*

(A) 6%   (B) 38%

(C) 48%   (D) 62%

 End of Practice Test 4 

Great job finishing the test!

 My Score

I got _____________ out of 30 questions right.

Check your answers in the Answer Key at the back of the book

Review any questions you missed. That's how we learn!

📊 Check Your Score Online!

Visit **ViewMath Academy** to enter your answers and see which topics you need to review. You can also explore lessons, take quizzes, track your scores, and save your progress!

viewmath.com/score/6.1.VA.19

Or go to viewmath.com/score and enter code: 6.1.VA.19

Practice Test 5

 30 Questions

✏️ Before You Start ✏️

- ✓ **Read each question carefully** before choosing your answer.
- ✓ **Show your work** on scratch paper when you need to.
- ✓ **Skip hard questions** and come back to them later.
- ✓ **Check your answers** when you're done.
- ✓ **Take your time** — there's no rush!

 You've Got This! 

Do your best and show what you know!

1. A class votes on two activities: hiking and swimming. The ratio of votes for hiking to swimming is $3:2$. There are 25 votes total. How many voted for swimming?

2. Which pair of ratios are equivalent?

(A) $2:3$ and $4:9$

(B) $2:3$ and $6:9$

(C) $2:3$ and $8:9$

(D) $2:3$ and $3:2$

3. A graph of equivalent ratios passes through $(0,0)$ and $(3,2)$. Which statement is true?

(A) The point $(9,4)$ is on the line.

(B) The point $(6,4)$ is on the line.

(C) The point $(6,5)$ is on the line.

(D) The point $(9,8)$ is on the line.

4. The table shows the original prices and discount percents for three items.

Item	Original Price	Discount
Backpack	$40	15%
Shoes	$60	10%
Hat	$20	25%

Which item has the greatest dollar amount of savings?

(A) Backpack ($6.00)

(B) Shoes ($6.00)

(C) Hat ($5.00)

(D) Backpack and shoes are tied.

5. What is $\dfrac{5}{6} \div \dfrac{1}{3}$?

(A) $\dfrac{5}{2}$

(B) $\dfrac{5}{18}$

(C) $\dfrac{2}{5}$

(D) $\dfrac{5}{3}$

6. A book costs \$12.95 and a bookmark costs \$3.48. Jenna buys one of each. How much does she spend in all?

Your Answer:

7. What is the GCF of 36 and 48?

(A) 12

(B) 6

(C) 24

(D) 144

8. A store sells shirts for \$24 and hats for \$16. Jake buys one of each. Which expression shows the total cost factored using the GCF?

(A) $4(6 + 4)$

(B) $8(3 + 2)$

(C) $2(12 + 8)$

(D) $4(6 + 16)$

9. Point A is at $(-5, 2)$. Point B is the reflection of Point A across the y-axis. What are the coordinates of Point B?

(A) $(5, 2)$

(B) $(-5, -2)$

(C) $(5, -2)$

(D) $(2, -5)$

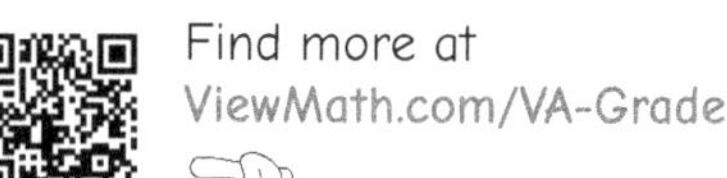

10. *Point P is at $(-2, y)$ and point Q is at $(-2, 5)$. The distance between P and Q is 8 units. Find the two possible values of y.*

Your Answer

11. *List all the terms in the expression $6a - 4b + 11$.*

Your Answer

12. *Evaluate $5(m - 3)$ when $m = 7$.*

A) 10

B) 20

C) 32

D) 17

13. *Simplify: $2(4m + 1) + 3m$*

Your Answer

14. *A tank holds 200 gallons of water and leaks 5 gallons per day. Which expression shows the water remaining after d days?*

A) $200 + 5d$

B) $200 - 5d$

C) $200d - 5$

D) $5d - 200$

15. *Is $x = 5$ a solution to $x > 5$?*

A) Yes, because 5 is equal to 5

B) Yes, because 5 is positive

C) No, because 5 is not greater than 5

D) No, because 5 is greater than 5

Find more at
ViewMath.com/VA-Grade6

16. Is $x = 3$ a solution to $x \leq 3$? Is $x = 3$ a solution to $x < 3$? Explain.

17. A rectangular piece of paper is 8 in by 6 in. If you cut it along a diagonal, what is the area of each triangular piece?

A) $14\ in^2$

B) $24\ in^2$

C) $48\ in^2$

D) $28\ in^2$

18. Which formula is used to find the area of a parallelogram?

A) $A = \frac{1}{2} \times b \times h$

B) $A = b \times h$

C) $A = \frac{1}{2}(b_1 + b_2) \times h$

D) $A = 2b + 2h$

19. A toy box is 3 ft long, 2 ft wide, and 2 ft tall. What is the volume?

A) $7\ ft^3$

B) $12\ ft^3$

C) $6\ ft^3$

D) $24\ ft^3$

Find more at
ViewMath.com/VA-Grade6

20. *Points $(1, 4)$ and $(1, -2)$ form a vertical segment. What is the length?*

  (A) 2 units   (B) 4 units

  (C) 5 units   (D) 6 units

21. *An L-shaped figure has vertices $(0, 0)$, $(10, 0)$, $(10, 4)$, $(6, 4)$, $(6, 8)$, and $(0, 8)$. What is the total area?*

Your Answer

22. *A rectangular prism has dimensions 10 m, 3 m, and 7 m. What is its surface area?*

  (A) $210\ m^2$ (B) $242\ m^2$

  (C) $262\ m^2$ (D) $121\ m^2$

23. *Point $K(-5, -2)$ is translated 7 units right and 4 units up. What are the coordinates of K'?*

  (A) $(2, 2)$ (B) $(-12, -6)$

  (C) $(2, -6)$ (D) $(-12, 2)$

24. *What does the number π (pi) represent?*

  (A) The radius divided by the diameter   (B) The ratio of a circle's circumference to its diameter

  (C) The area of any circle with radius 1   (D) The diameter divided by the circumference

25. *Which measure of center is more affected by an outlier?*

  (A) Median   (B) Mean

  (C) Both are equally affected   (D) Neither is affected

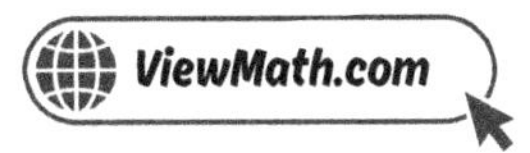

26. *Data:* $20, 22, 24, 26, 28$. *The mean is 24. What is the MAD?*

(A) 2
(B) 2.4

(C) 4
(D) 8

27. *A frequency table shows students' favorite lunch: Pizza (12), Chicken (8), Salad (4), Burgers (6). Find the total and the mode.*

Your Answer:

28. *Store A daily sales: median $= \$500$, range $= \$200$. Store B daily sales: median $= \$500$, range $= \$800$. What can you conclude?*

(A) *Both stores are equally consistent.*
(B) *Store A is more consistent in its daily sales.*

(C) *Store B is more consistent in its daily sales.*
(D) *Store B has lower typical sales.*

29. *The spinner below has 8 equal sections. Find each probability as a fraction in simplest form.*
(a) P(red)   (b) P(blue)   (c) P(not green)

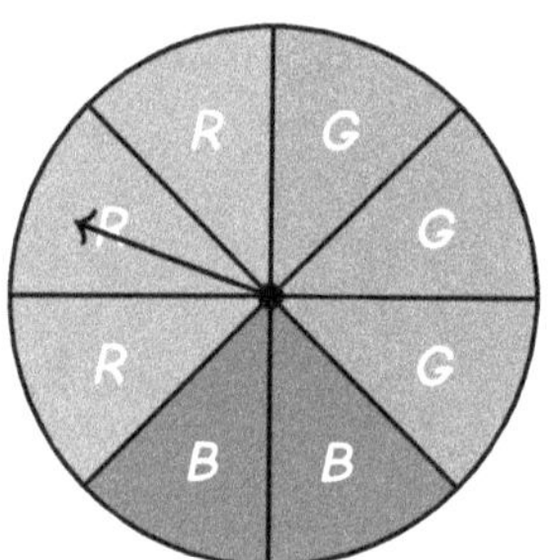

30. *A school has 600 students. A circle graph shows: Grade 6 is 35%, Grade 7 is 40%, Grade 8 is 25%. How many more students are in Grade 7 than in Grade 8?*

(A) 15

(B) 40

(C) 90

(D) 150

 # End of Practice Test 5 

Great job finishing the test!

☑ My Score

I got _____________ out of 30 questions right.

Check your answers in the Answer Key at the back of the book.

Review any questions you missed. That's how we learn!

📊 Check Your Score Online!

Visit **ViewMath Academy** to enter your answers and see which topics you need to review. You can also explore lessons, take quizzes, track your scores, and save your progress!

viewmath.com/score/6.1.VA.20

Or go to viewmath.com/score and enter code: 6.1.VA.20

Practice Test 6

30 Questions

✏ Before You Start ✏

- ✓ **Read each question carefully** before choosing your answer.
- ✓ **Show your work** on scratch paper when you need to.
- ✓ **Skip hard questions** and come back to them later.
- ✓ **Check your answers** when you're done.
- ✓ **Take your time** — there's no rush!

⭐ You've Got This! ⭐

Do your best and show what you know!

1. A smoothie recipe uses 3 bananas **for every** 4 cups of yogurt. Write the ratio of bananas to yogurt. Then write the ratio of yogurt to bananas.

2. Look at this ratio table. What is the missing value?

Apples	Oranges
3	5
6	?

(A) 8

(B) 10

(C) 15

(D) 11

3. Which table matches a graph that passes through $(0,0)$, $(2,3)$, and $(4,6)$?

(A)

x	y
2	3
6	9

(B)

x	y
2	3
6	8

(C)

x	y
3	2
6	9

(D)

x	y
2	4
6	12

4. What is 10% of 350?

(A) 3.5

(B) 35

(C) 350

(D) 70

Get Online

Find more at
ViewMath.com/VA-Grade6

5. Maya jogged $\frac{7}{8}$ of a mile. Each lap around the park is $\frac{1}{4}$ of a mile. How many laps did she complete?

(A) $\frac{7}{32}$

(B) 4

(C) 2

(D) $3\frac{1}{2}$

6. What is $4.56 \div 0.6$?

(A) 0.76

(B) 76

(C) 7.6

(D) 0.076

7. Bus A arrives every 8 minutes and Bus B arrives every 10 minutes. Both buses arrive at noon. After how many minutes will both buses arrive at the same time again?

(A) 2

(B) 18

(C) 80

(D) 40

8. Which shows $12 + 28$ written as a product using the GCF?

(A) $2(6 + 14)$

(B) $4(3 + 7)$

(C) $4(3 + 28)$

(D) $7(2 + 4)$

9. In which quadrant is the point $(-3, -6)$ located?

(A) Quadrant I

(B) Quadrant II

(C) Quadrant III

(D) Quadrant IV

Find more at
ViewMath.com/VA-Grade6

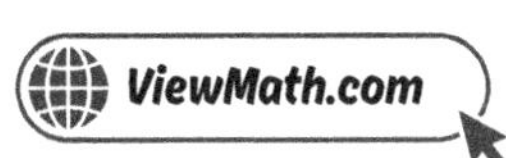

10. A bird sits at $(1, -2)$ on a coordinate grid. It flies straight up to $(1, 5)$. How many units did it fly?

(A) 3

(B) -7

(C) 7

(D) 12

11. Which statement is true about the expression $x + 5$?

(A) The coefficient of x is 0

(B) There are 3 terms

(C) 5 is the coefficient of x

(D) The coefficient of x is 1

12. The input-output table below uses the expression $2x + 3$. Which output value is incorrect?

Input (x)	Output ($2x + 3$)
1	5
2	7
3	8
4	11

(A) $x = 1$

(B) $x = 2$

(C) $x = 3$

(D) $x = 4$

13. A student simplified $4(2x + 3) = 8x + 3$. Find and fix the error.

14. In the expression $12p + 6$, p stands for the number of people. What does 12 represent?

(A) A flat fee

(B) The cost per person

(C) The number of people

(D) The total cost

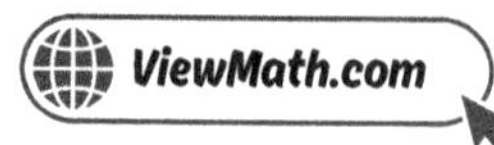

15. *Which of the following is a solution to $n \leq 8$?*

(A) $n = 9$

(B) $n = 8.5$

(C) $n = 8$

(D) $n = 10$

16. *Describe the graph of $n \leq 5$.*

(A) Open circle at 5, shade left

(B) Closed circle at 5, shade left

(C) Open circle at 5, shade right

(D) Closed circle at 5, shade right

17. *A right triangle has legs of 9 in and 12 in. What is its area?*

18. *A table top is shaped like a trapezoid. The parallel sides measure 2 ft and 4 ft, and the distance between them is 3 ft. What is the area of the table top?*

Your Answer:

19. *A rectangular prism is $2\frac{1}{2}$ ft long, 4 ft wide, and 3 ft tall. What is its volume?*

(A) 24 ft³

(B) 30 ft³

(C) 9.5 ft³

(D) 20 ft³

20. A segment goes from $(0, -4)$ to $(0, 3)$. If both endpoints are reflected across the y-axis, what are the new endpoints?

21. A rectangle on the coordinate plane has an area of 56 square units. Its length is 8 units. What is the width?

(A) 6 units

(B) 7 units

(C) 8 units

(D) 48 units

22. A box is 15 in long, 6 in wide, and 4 in tall. How many square inches of cardboard are needed to make the box?

23. Point $M(-2, 7)$ moves to $M'(4, 7)$. What transformation occurred?

(A) Reflection across the x-axis

(B) Reflection across the y-axis

(C) Translation 6 units right

(D) Translation 6 units left

24. A circle has a diameter of 12 m. What is its area? Use $\pi \approx 3.14$.

(A) $37.68 \ m^2$

(B) $113.04 \ m^2$

(C) $226.08 \ m^2$

(D) $452.16 \ m^2$

25. The mean of five numbers is 20. What is the sum of the five numbers?

(A) 4

(B) 25

(C) 80

(D) 100

Find more at
ViewMath.com/VA-Grade6

26. Team A scores: $60, 62, 64, 66, 68$. Team B scores: $40, 50, 64, 78, 88$. Both have the same mean. Which team is more consistent?

(A) Team A, because its range is smaller.

(B) Team B, because its range is smaller.

(C) They are equally consistent because their means are equal.

(D) Neither is consistent.

27. In a histogram, there are no gaps between the bars because —

(A) the data values are all the same

(B) the intervals are consecutive and cover every value in the range

(C) the bars overlap

(D) it shows categorical data

28. Daily steps for Week 1: median $= 8,000$, IQR $= 2,000$. Week 2: median $= 9,500$, IQR $= 1,200$. Which week shows a higher typical step count and more consistent activity?

(A) Week 1 for both

(B) Week 2 for both

(C) Week 1 higher, Week 2 more consistent

(D) Week 2 higher, Week 1 more consistent

29. A fair coin is flipped once. What is the probability of landing on heads?

(A) $\dfrac{1}{4}$

(B) $\dfrac{1}{3}$

(C) $\dfrac{1}{2}$

(D) 1

30. A circle graph shows how 200 students get to school: Bus 40%, Car 25%, Walk 20%, Bike 15%. How many students take the bus?

(A) 30

(B) 60

(C) 80

(D) 100

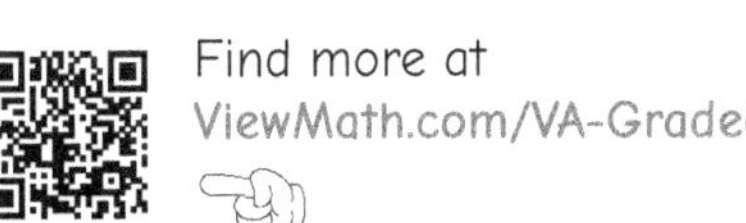
Find more at
ViewMath.com/VA-Grade6

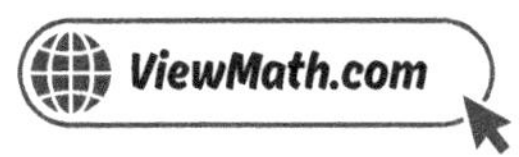

 *End of Practice Test 6* 

Great job finishing the test!

 My Score

I got ____________ out of 30 questions right.

Check your answers in the Answer Key at the back of the book

Review any questions you missed. That's how we learn!

Check Your Score Online!

Visit **ViewMath Academy** to enter your answers and see which topics
you need to review. You can also explore lessons, take quizzes, track
your scores, and save your progress!

viewmath.com/score/6.1.VA.21

Or go to viewmath.com/score and enter code: 6.1.VA.21

Practice Test 7

 30 Questions

 Before You Start 

- ✓ **Read each question carefully** before choosing your answer.
- ✓ **Show your work** on scratch paper when you need to.
- ✓ **Skip hard questions** and come back to them later.
- ✓ **Check your answers** when you're done.
- ✓ **Take your time** — there's no rush!

 You've Got This! 

Do your best and show what you know!

1. A lemonade stand sells 7 cups of lemonade **for each** 2 cups of iced tea. If they sold 21 cups of lemonade, how many cups of iced tea did they sell?

Your Answer

2. A ratio table for pens to notebooks starts with $2 : 3$. Which of these is a valid row?

(A) $4 : 5$

(B) $8 : 12$

(C) $6 : 12$

(D) $10 : 12$

3. A ratio graph passes through $(4, 10)$ and the origin. List two other points on this line.

Your Answer

4. A town has 4,000 residents. 15% are children. How many children live in the town?

(A) 60

(B) 600

(C) 150

(D) 1,500

5. You have $\frac{5}{6}$ of a gallon of paint. Each birdhouse needs $\frac{1}{6}$ of a gallon. How many birdhouses can you paint?

6. What is $4.36 + 9.7$?

(A) 13.06

(B) 14.06

(C) 14.6

(D) 5.06

7. Sam visits the library every 9 days and Mia visits the library every 12 days. They are both at the library today. In how many days will they be at the library on the same day again?

8. A student rewrote $28 + 42$ as $7(4 + 8)$. What error did the student make?

(A) 7 is not a factor of 28

(B) $42 \div 7$ equals 6, not 8

(C) $28 \div 7$ equals 3, not 4

(D) The distributive property cannot be used with addition

9. In which quadrant is the point $(5, 3)$ located?

(A) Quadrant I

(B) Quadrant II

(C) Quadrant III

(D) Quadrant IV

10. Two traffic lights are at $(-3, 5)$ and $(4, 5)$ along the same street on a city grid. How many blocks apart are they?

(A) 1

(B) 5

(C) 7

(D) 9

11. Write an expression that has 3 terms, a coefficient of 4, and a constant of 10.

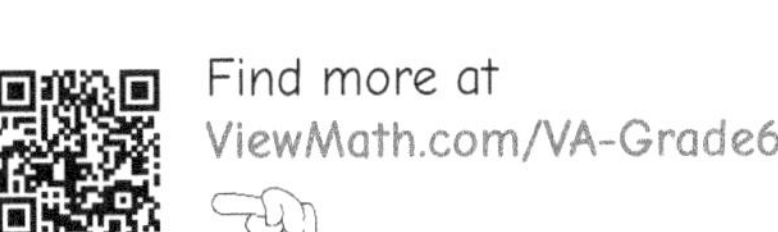

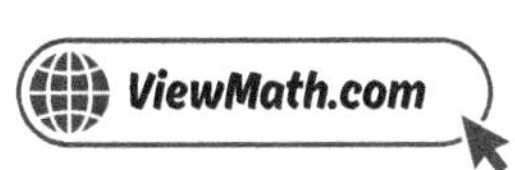

12. Evaluate w^3 when $w = 2$.

A 6

B 8

C 16

D 32

13. Which pair of expressions are NOT equivalent?

A $3x + 5x$ and $8x$

B $2(a + 4)$ and $2a + 8$

C $6y - 2y$ and $4y$

D $3x + 2x$ and $5x^2$

14. A plumber charges \$50 for a house visit plus \$40 per hour of work. Which of the following represents the cost of a 3-hour job?

A \$90

B \$120

C \$170

D \$190

15. Write an inequality: You must be no fewer than 16 years old to drive. Let a = age.

Your Answer:

16. When graphing $x \leq 6$ on a number line, which direction do you shade?

A To the right

B To the left

C Both directions

D No shading needed

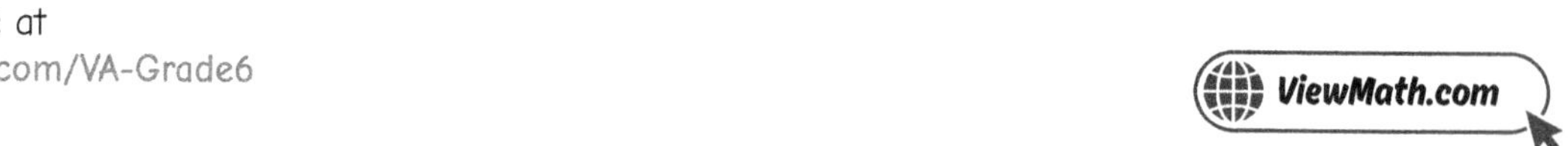

17. What is the area of a triangle with base 10 cm and height 6 cm?

(A) $60 \ cm^2$

(B) $30 \ cm^2$

(C) $16 \ cm^2$

(D) $32 \ cm^2$

18. A trapezoid has bases 9 in and 15 in and height 6 in. What is the area?

(A) $90 \ in^2$

(B) $72 \ in^2$

(C) $45 \ in^2$

(D) $135 \ in^2$

19. A fish tank is 20 in long, 10 in wide, and 12 in tall. What is the volume of the tank?

(A) $42 \ in^3$

(B) $240 \ in^3$

(C) $1,200 \ in^3$

(D) $2,400 \ in^3$

20. A triangle has vertices $P(2, 4)$, $Q(6, 4)$, and $R(6, 1)$. Reflect the triangle across the y-axis. What are the coordinates of P'?

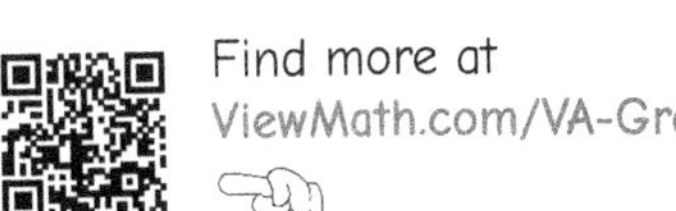

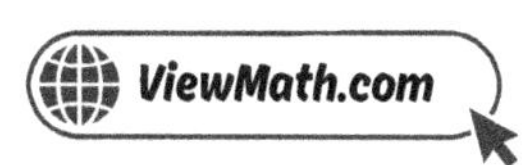

21. An L-shaped figure has vertices $(0,0)$, $(8,0)$, $(8,3)$, $(4,3)$, $(4,7)$, and $(0,7)$. A student splits it into two rectangles. Which pair of rectangles correctly covers the figure?

(A) 8×7 and 4×3

(B) 8×3 and 4×4

(C) 4×7 and 4×3

(D) 8×3 and 4×7

22. A rectangular prism has dimensions 4 cm by 4 cm by 9 cm. What is the surface area?

(A) $144\ cm^2$

(B) $176\ cm^2$

(C) $160\ cm^2$

(D) $208\ cm^2$

23. Reflect the point $(8, -3)$ across the x-axis. What are the new coordinates?

24. Two circles are shown below. Circle P has a radius of 4 in and Circle Q has a radius of 8 in. The area of Circle Q is how many times the area of Circle P?

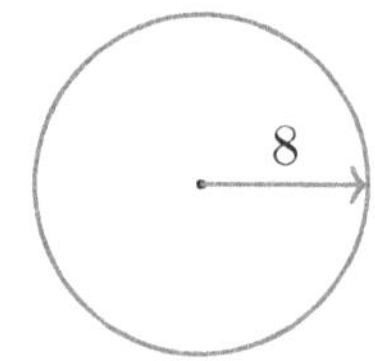

(A) 2 times

(B) 3 times

(C) 4 times

(D) 8 times

25. Data: $16, 20, 24, 28, 32, 36$. Find the median.

26. *Which statement about MAD is true?*

(A) *MAD measures the middle value of the data.* (B) *MAD tells you the average distance of each value from the mean.*

(C) *MAD is always larger than the range.* (D) *MAD equals the range divided by 2.*

27. *Look at the histogram below showing student test scores.*

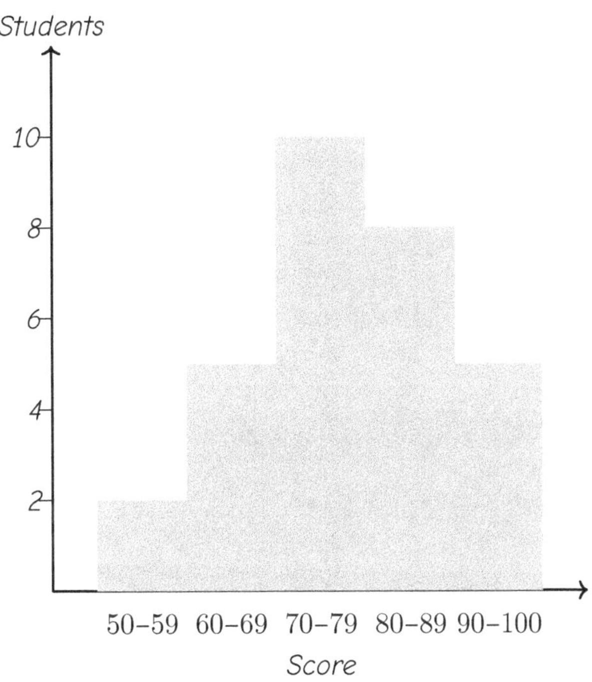

How many students took the test? What is the modal interval? What fraction of students scored 70 or higher?

Your Answer:

28. *Which two measures together best describe the "center" and "spread" of a data set?*

(A) *Mean and mode* (B) *Median and IQR*

(C) *Range and maximum* (D) *Minimum and maximum*

29. Event A has a probability of $\frac{2}{5}$ and Event B has a probability of $\frac{3}{4}$. Which statement is true?

   (A) Event A is more likely than Event B

   (B) Event B is more likely than Event A

   (C) Both events are equally likely

   (D) Neither event can happen

30. A sector in a circle graph represents 75% of the data. Which fraction in simplest form equals 75%?

   (A) $\frac{1}{4}$

   (B) $\frac{3}{5}$

   (C) $\frac{3}{4}$

   (D) $\frac{7}{10}$

 End of Practice Test 7 

Great job finishing the test!

My Score

I got _____________ out of 30 questions right.

Check your answers in the Answer Key at the back of the book.

Review any questions you missed. That's how we learn!

📊 Check Your Score Online!

Visit **ViewMath Academy** to enter your answers and see which topics you need to review. You can also explore lessons, take quizzes, track your scores, and save your progress!

viewmath.com/score/6.1.VA.22

Or go to *viewmath.com/score* and enter code: 6.1.VA.22

Practice Test 8

📋 30 Questions

✏️ Before You Start ✏️

- ✓ **Read each question carefully** before choosing your answer.
- ✓ **Show your work** on scratch paper when you need to.
- ✓ **Skip hard questions** and come back to them later.
- ✓ **Check your answers** when you're done.
- ✓ **Take your time** — there's no rush!

⭐ You've Got This! ⭐

Do your best and show what you know!

1. The ratio of cats to dogs at a pet store is $3 : 4$. Each part in the tape diagram represents 2 animals. How many dogs are there?

(A) 6

(B) 8

(C) 3

(D) 4

2. The graph below shows points that represent equivalent ratios of flour to sugar in a recipe.

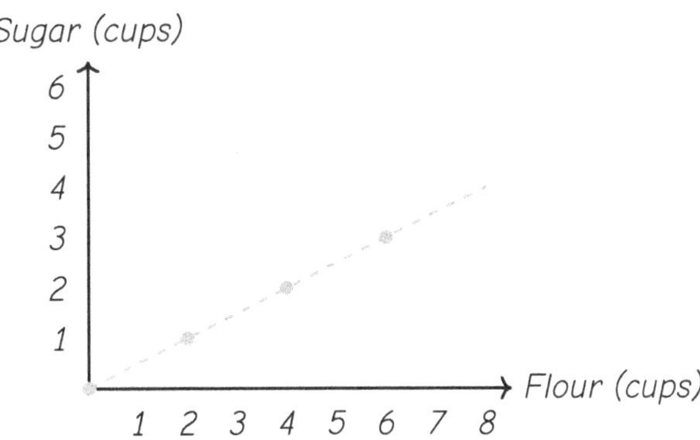

Part A: What is the ratio of flour to sugar?

Part B: If you need 8 cups of flour, how many cups of sugar do you need?

3. True or false: a graph of equivalent ratios can be a curved line. Explain.

4. A class of 30 students was surveyed. 12 chose chocolate ice cream. What percent chose chocolate?

(A) 12%

(B) 30%

(C) 40%

(D) 60%

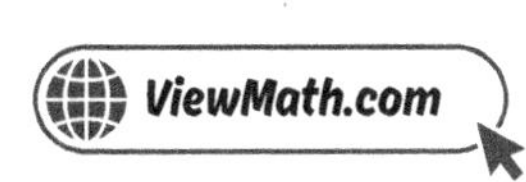

5. *Evaluate:* $\dfrac{9}{10} \div \dfrac{3}{5}$

6. *A student multiplied* 1.4×0.3 *and showed the following work.*

Step 1: *Multiply as whole numbers:* $14 \times 3 = 42$.

Step 2: *Count* 1 *decimal place.*

Answer: 4.2

What is the **correct** *product?*

A) 4.2

B) 42

C) 0.42

D) 0.042

7. *What is the GCF of* 54 *and* 90?

A) 9

B) 6

C) 270

D) 18

8. *Mia bought 7 packs of pencils and 7 packs of erasers. Each pack of pencils costs $3 and each pack of erasers costs $2. Which expression uses the distributive property to find the total cost?*

A) $7 + 3 + 2$

B) $7 \times 3 \times 2$

C) $7(3 + 2)$

D) $(7 + 3)(7 + 2)$

Find more at
ViewMath.com/VA-Grade6

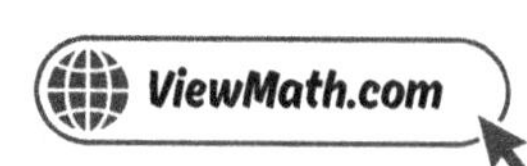

9. If the point $(3, 5)$ is reflected across the y-axis, what are the coordinates of its image?

   (A) $(-3, 5)$ (B) $(3, -5)$

   (C) $(-3, -5)$ (D) $(5, 3)$

10. **Part A:** Find the horizontal distance between $(-5, 3)$ and $(2, 3)$.

Part B: Find the vertical distance between $(2, 3)$ and $(2, -4)$.

Part C: If you walk from $(-5, 3)$ to $(2, 3)$ and then from $(2, 3)$ to $(2, -4)$, what is the total distance you walk?

11. The table below shows different expressions. Which row has the coefficient identified incorrectly?

Row	Expression	Coefficient of the variable
1	$8p + 2$	8
2	$r - 5$	0
3	$3(x + 4)$	3
4	$10 + 6y$	6

   (A) Row 1   (B) Row 2

   (C) Row 3   (D) Row 4

12. Evaluate $\dfrac{n}{3} + 8$ when $n = 15$.

   (A) 5   (B) 11

   (C) 13   (D) 23

13. *Simplify:* $3a + 7 + 5a - 2a + 3$

14. The chart shows how a savings account grows. Write an expression using w (weeks) that matches the pattern, and predict the amount after 10 weeks.

Week (w)	Savings ($)
0	$50
1	$65
2	$80
3	$95

15. Write a real-world situation that can be described by $x \leq 15$.

16. Write an inequality whose graph has an open circle at 12 and shading to the left.

17. A triangle has an area of $45\ m^2$ and a base of $10\ m$. What is the height?

18. A park is shaped like a trapezoid with parallel sides of 40 m and 60 m and a height of 30 m. What is the area of the park?

(A) 1,800 m^2 (B) 1,200 m^2

(C) 2,400 m^2 (D) 1,500 m^2

19. A box is 2.5 cm long, 4 cm wide, and 6 cm tall. What is the volume?

20. The rectangle $ABCD$ is shown below. After reflecting across the x-axis, in which quadrant will vertex A' be located?

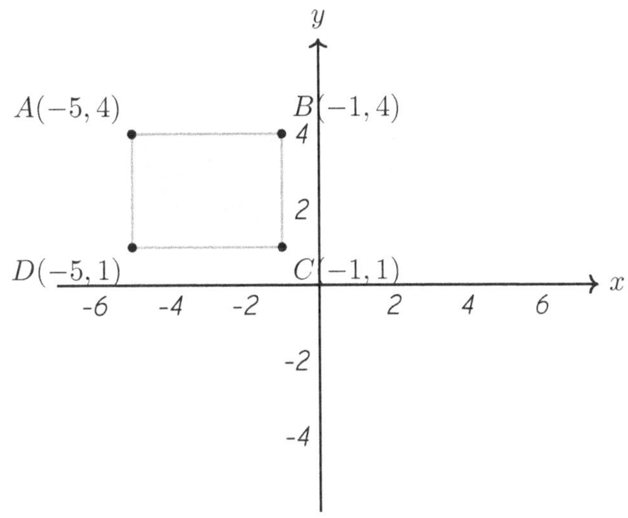

(A) Quadrant I   (B) Quadrant II

(C) Quadrant III   (D) Quadrant IV

21. A right triangle has vertices $(-3, -2)$, $(5, -2)$, and $(-3, 4)$. What is the area?

Your Answer:

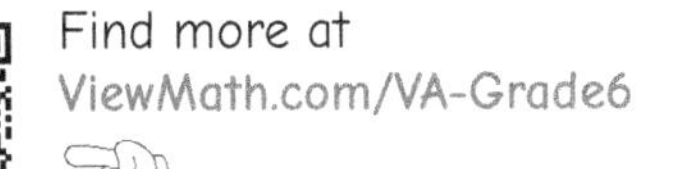
Find more at
ViewMath.com/VA-Grade6

22. A cube has edges of 5 in. What is the surface area?

   (A) 25 in^2 (B) 125 in^2

   (C) 150 in^2 (D) 75 in^2

23. What is the reflection of $(-4, 2)$ across the y-axis?

   (A) $(-4, -2)$ (B) $(4, -2)$

   (C) $(4, 2)$ (D) $(2, -4)$

24. A circle has a diameter of 18 inches. What is its area? Use $\pi \approx 3.14$.

Your Answer

25. Data: $3, 4, 5, 6, 7, 8, 60$. Find the mean and median. Which better describes a typical value?

26. Which measure of spread is most affected by outliers?

   (A) IQR   (B) MAD

   (C) Range   (D) Median

27. A dot plot of daily steps: 3000 (1), 4000 (2), 5000 (6), 6000 (4), 7000 (2). What is the mode? What is the range?

Your Answer

28. Which pair of measures should you report when outliers are present?

(A) Mean and range

(B) Mode and maximum

(C) Median and IQR

(D) Mean and IQR

29. A bag contains 4 green marbles, 6 red marbles, and 10 blue marbles. What is the probability of randomly drawing a green marble? Write your answer as a percent.

30. A circle graph shows that 15% of all donations went to animal shelters. If animal shelters received $450, what was the total amount of donations?

(A) $1,500

(B) $3,000

(C) $4,500

(D) $30,000

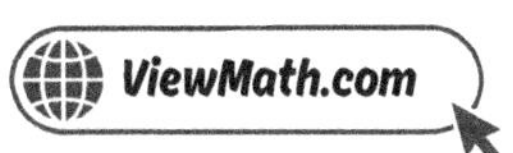

 *End of Practice Test 8* 

Great job finishing the test!

 *My Score*

I got _____________ out of 30 questions right.

Check your answers in the Answer Key at the back of the book.

Review any questions you missed. That's how we learn!

📊 Check Your Score Online!

Visit **ViewMath Academy** to enter your answers and see which topics you need to review. You can also explore lessons, take quizzes, track your scores, and save your progress!

viewmath.com/score/6.1.VA.23

Or go to viewmath.com/score and enter code: 6.1.VA.23

Practice Test 9

☑ 30 Questions

✏ Before You Start ✏

✔ **Read each question carefully** before choosing your answer.

✔ **Show your work** on scratch paper when you need to.

✔ **Skip hard questions** and come back to them later.

✔ **Check your answers** when you're done.

✔ **Take your time** — there's no rush!

⭐ *You've Got This!* ⭐

Do your best and show what you know!

1. The ratio of red to blue to green beads is $1:3:2$. There are 18 beads total. How many blue beads are there?

   (A) 3

   (B) 6

   (C) 9

   (D) 12

2. The ratio of red to blue beads is $1:4$. If there are 3 red beads, how many blue beads are there?

   (A) 4

   (B) 7

   (C) 12

   (D) 8

3. Two ratio graphs both pass through the origin. Line A goes through $(1, 4)$ and Line B goes through $(1, 3)$. Which line is steeper?

   (A) Line A

   (B) Line B

   (C) They have the same steepness.

   (D) Cannot be determined.

4. A jacket costs $80. It is 25% off. How much do you save?

   (A) $25

   (B) $20

   (C) $60

   (D) $15

5. What is $\dfrac{4}{9} \div \dfrac{2}{3}$?

   (A) $\dfrac{2}{3}$

   (B) $\dfrac{8}{27}$

   (C) $\dfrac{3}{2}$

   (D) $\dfrac{2}{9}$

6. A notebook costs $3.49 and a pen costs $1.75. How much do they cost together?

A) $5.14

B) $5.24

C) $4.24

D) $5.34

7. What is the GCF of 16 and 40?

A) 4

B) 80

C) 8

D) 16

8. A baker makes 54 chocolate cookies and 36 vanilla cookies. The table below summarizes the cookies.

Cookie Type	Quantity
Chocolate	54
Vanilla	36

She wants to pack the cookies into identical gift bags with no cookies left over.

Part A: Find the GCF of 54 and 36.

Part B: Use the distributive property to write $54 + 36$ as a product of the GCF and a sum.

Part C: How many gift bags can she make, and how many of each type of cookie will be in each bag?

9. What are the coordinates of the origin?

A) $(1, 1)$

B) $(0, 1)$

C) $(1, 0)$

D) $(0, 0)$

10. Three points are plotted on the coordinate plane below.

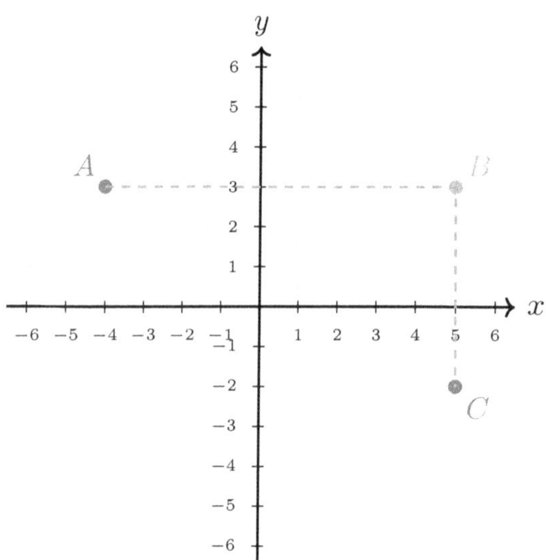

Part A: *What is the distance from A to B?*

Part B: *What is the distance from B to C?*

Part C: *If you walk from A to B and then from B to C, what is the total distance?*

11. What is the coefficient of d in the expression $15 - d + 3$?

- (A) 0
- (B) 1
- (C) −1
- (D) 15

12. Evaluate $3x + 7$ when $x = 4$.

- (A) 19
- (B) 34
- (C) 15
- (D) 47

13. *Which expression is equivalent to* $10a - 2a + 7$?

(A) $12a + 7$

(B) $8a + 7$

(C) $8a - 7$

(D) $15a$

14. *A rectangle has length l and width 5. Which expression represents its area?*

(A) $l + 5$

(B) $2l + 10$

(C) $5l$

(D) $l - 5$

15. *Which phrase matches the inequality $t \leq 30$?*

(A) *The temperature is more than 30 degrees*

(B) *The temperature is exactly 30 degrees*

(C) *The temperature is no more than 30 degrees*

(D) *The temperature is at least 30 degrees*

16. *Which inequality has a graph with an open circle at 0 and shading to the right?*

(A) $x \leq 0$

(B) $x \geq 0$

(C) $x > 0$

(D) $x < 0$

17. *A triangle has base 3.5 cm and height 4 cm. What is its area?*

(A) $14 \ cm^2$

(B) $7.5 \ cm^2$

(C) $7 \ cm^2$

(D) $15 \ cm^2$

18. *A parallelogram has base 12 m and a slant side of 8 m. The height is 6 m. What is the area?*

(A) $96 \ m^2$

(B) $72 \ m^2$

(C) $48 \ m^2$

(D) $36 \ m^2$

Find more at
ViewMath.com/VA-Grade6

19. A rectangular prism has length $\frac{1}{2}$ m, width $\frac{1}{2}$ m, and height $\frac{1}{2}$ m. What is the volume?

(A) $\frac{1}{8}$ m^3

(B) $\frac{1}{4}$ m^3

(C) $\frac{3}{2}$ m^3

(D) $\frac{1}{2}$ m^3

20. Sara reflects the point $(3, -8)$ across the y-axis and gets $(-3, 8)$. Is she correct?

(A) Yes, she is correct.

(B) No, the correct answer is $(-3, -8)$.

(C) No, the correct answer is $(3, 8)$.

(D) No, the correct answer is $(-8, 3)$.

21. A right triangle has vertices $(1, 1)$, $(1, 9)$, and $(5, 1)$. What is the area?

(A) 32 square units

(B) 16 square units

(C) 20 square units

(D) 8 square units

22. What is the surface area of a rectangular prism with length 6 cm, width 4 cm, and height 3 cm?

(A) 72 cm^2

(B) 108 cm^2

(C) 54 cm^2

(D) 96 cm^2

23. Point $A(3, -1)$ moves to $A'(3, 5)$. Which translation describes this move?

(A) 6 units right

(B) 6 units up

(C) 4 units left

(D) 4 units down

24. A circle has a diameter of 10 cm. What is its circumference? Use $\pi \approx 3.14$.

(A) 15.7 cm

(B) 31.4 cm

(C) 62.8 cm

(D) 314 cm

Find more at
ViewMath.com/VA-Grade6

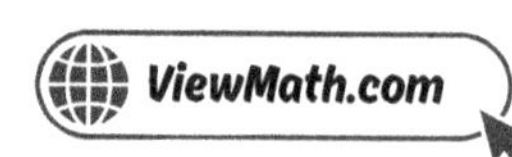

25. Look at the dot plot below showing the number of goals scored per game.

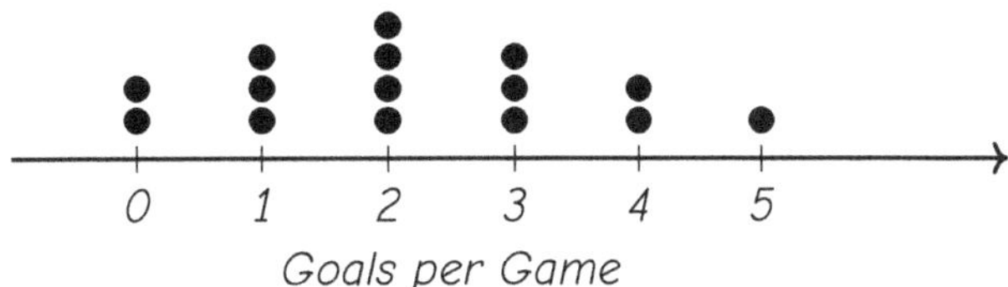

What is the median number of goals?

(A) 1

(B) 2

(C) 2.5

(D) 3

26. If every value in a data set is increased by 5, what happens to the range?

(A) It increases by 5.

(B) It decreases by 5.

(C) It stays the same.

(D) It doubles.

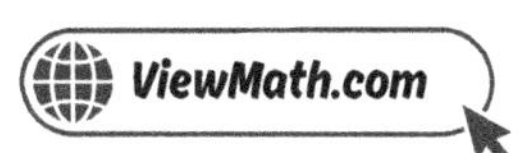

27. The histogram below shows the number of pages students read last week.

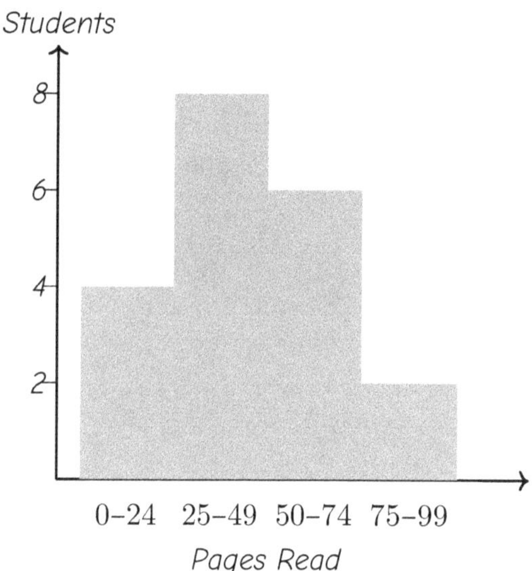

Which interval is the modal interval (contains the most students)?

(A) 0–24

(B) 25–49

(C) 50–74

(D) 75–99

28. Which measure of spread is affected by every single data value?

(A) IQR

(B) Range

(C) Median

(D) Mode

29. A jar holds 10 jellybeans: 4 cherry, 4 grape, and 2 lemon. Which event is **equally likely** as drawing a cherry jellybean?

(A) Drawing a lemon jellybean

(B) Drawing a grape jellybean

(C) Not drawing a jellybean

(D) Drawing a cherry or lemon jellybean

30. In a circle graph of 300 students, 40% chose basketball, 35% chose soccer, and the rest chose swimming. How many students chose swimming?

Your Answer

 *End of Practice Test 9* 

Great job finishing the test!

✅ My Score

I got __________ out of 30 questions right.

Check your answers in the Answer Key at the back of the book.

💡 Review any questions you missed. That's how we learn!

📊 Check Your Score Online!

*Visit **ViewMath Academy** to enter your answers and see which topics you need to review. You can also explore lessons, take quizzes, track your scores, and save your progress!*

viewmath.com/score/6.1.VA.24

Or go to viewmath.com/score and enter code: 6.1.VA.24

10

Practice Test 10

 30 Questions

✏️ Before You Start ✏️

- ✓ **Read each question carefully** before choosing your answer.
- ✓ **Show your work** on scratch paper when you need to.
- ✓ **Skip hard questions** and come back to them later.
- ✓ **Check your answers** when you're done.
- ✓ **Take your time** — there's no rush!

⭐ You've Got This! ⭐

Do your best and show what you know!

1. The tape diagram below shows the ratio of apple juice to orange juice in a drink.

Apple: ▢▢▢
Orange: ▢▢▢▢▢

If each part equals 4 ounces, how many total ounces of drink are there?

(A) 12

(B) 20

(C) 32

(D) 8

2. The ratio table below shows the relationship between bags of seed and the area of garden they cover.

Bags of Seed	Area (sq ft)
2	50
4	100
6	?
8	200

What is the missing value?

(A) 125

(B) 150

(C) 175

(D) 160

3. Juice costs \$2 per bottle. Which point would **not** appear on the graph of this ratio?

(A) $(1, 2)$

(B) $(3, 6)$

(C) $(4, 10)$

(D) $(5, 10)$

4. Maria saved \$45 toward a \$90 bike. What percent has she saved?

(A) 25%

(B) 45%

(C) 50%

(D) 75%

5. If $\dfrac{a}{b} \div \dfrac{c}{d} = \dfrac{a}{b} \times \dfrac{?}{?}$, what fraction replaces the question marks?

(A) $\dfrac{d}{c}$

(B) $\dfrac{c}{d}$

(C) $\dfrac{b}{a}$

(D) $\dfrac{a}{b}$

6. The table shows the distances a cyclist rode each day.

Day	Distance (km)
Monday	8.76
Tuesday	12.30
Wednesday	6.48

Part A: What is the total distance the cyclist rode over the three days?

Part B: If the cyclist wants to divide the total distance equally over the 3 days, how many kilometers per day is that?

Your Answer

7. Find the LCM of 5 and 7.

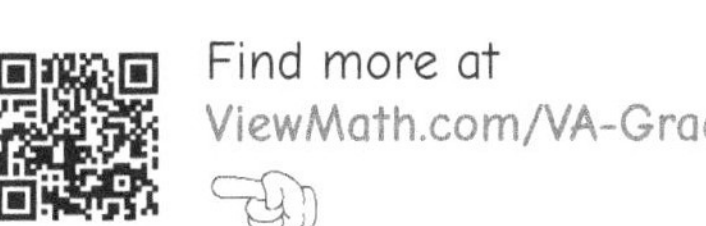

8. Rewrite $12 + 18 + 30$ using the GCF and a sum.

9. In which quadrant is the point $(2, -5)$ located?

(A) Quadrant I

(B) Quadrant II

(C) Quadrant III

(D) Quadrant IV

10. What is the distance between $(4, -6)$ and $(4, -1)$?

(A) -7

(B) -5

(C) 5

(D) 7

11. Look at the expression diagram below. Identify: (a) the coefficient of y, (b) the constant, and (c) the total number of terms.

$$9y \ + \ \underline{2x} \ \underline{-} \ \underline{5} \ \underline{+} \ \underline{y}$$

Term 1 Term 2 Term 3 Term 4

12. Evaluate $2(3y - 1)$ when $y = 5$.

(A) 28

(B) 14

(C) 29

(D) 10

13. A student's work simplifying $3(2x + 4) + x$ is shown below. Find and correct the error.

Step 1: $3(2x + 4) + x$

Step 2: $6x + 4 + x$

Step 3: $7x + 4$

14. You have 100 pages to read and read r pages each day. Which expression tells how many pages are left after 4 days?

(A) $100 + 4r$

(B) $100 - 4r$

(C) $4r - 100$

(D) $100r - 4$

15. Which of the following is true about the inequality $x > -2$?

(A) $x = -2$ is a solution

(B) $x = -3$ is a solution

(C) $x = 0$ is a solution

(D) $x = -2.5$ is a solution

16. Describe the graph of $x > 1$. What kind of circle? Which direction to shade?

17. A triangle has base 2.4 m and height 5 m. What is the area?

(A) $12 \ m^2$

(B) $7.4 \ m^2$

(C) $6 \ m^2$

(D) $24 \ m^2$

18. *Find the total area of the composite figure shown below. It is made of a rectangle and a trapezoid.*

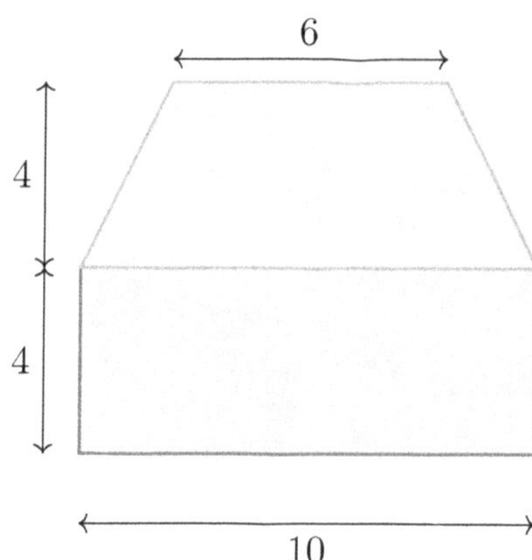

Your Answer:

19. *A shipping box has dimensions 12 in by 8 in by 6 in. How many 1-inch cubes would fit inside?*

(A) 26 cubes

(B) 96 cubes

(C) 576 cubes

(D) 288 cubes

20. *What is the reflection of $(-3, 2)$ across the y-axis?*

(A) $(-3, -2)$

(B) $(3, -2)$

(C) $(3, 2)$

(D) $(2, -3)$

21. *A rectangle has vertices $(0, 0)$, $(6, 0)$, $(6, 4)$, and $(0, 4)$. What is the area?*

(A) 10 square units

(B) 20 square units

(C) 24 square units

(D) 12 square units

22. Which pair of dimensions gives a rectangular prism with the greatest surface area?

(A) $2 \times 3 \times 10$ (B) $4 \times 4 \times 4$

(C) $1 \times 5 \times 8$ (D) $3 \times 3 \times 6$

23. Point $G(-3, 5)$ is reflected across the y-axis and then translated 2 units down. What are the final coordinates?

24. Find the area of the semicircle shown below. Use $\pi \approx 3.14$.

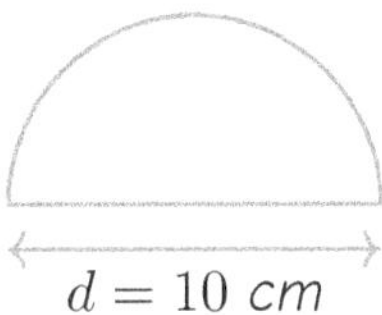

(A) $15.7\ cm^2$ (B) $39.25\ cm^2$

(C) $78.5\ cm^2$ (D) $157\ cm^2$

25. Four students' heights in inches are: $58, 60, 62, 64$. A fifth student who is 82 inches tall joins. What happens to the mean?

(A) It stays at 61.   (B) It increases to 65.2.

(C) It decreases.   (D) It increases to 82.

26. Two data sets have the same mean of 50. Data set A: $MAD = 2$. Data set B: $MAD = 10$. Which data set has more variability?

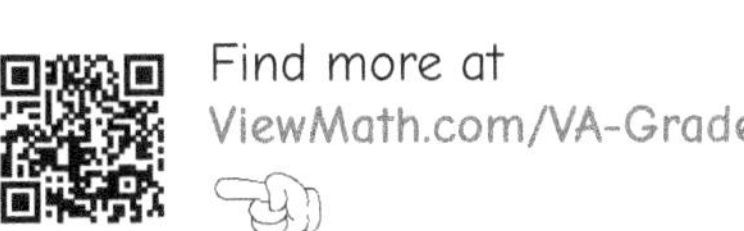

27. *Quiz scores:* $7, 8, 8, 9, 8, 10, 9, 7, 8, 9.$ *What is the mode?*

   (A) 7   (B) 8

   (C) 9   (D) 10

28. *Two box plots are shown on the same number line.*

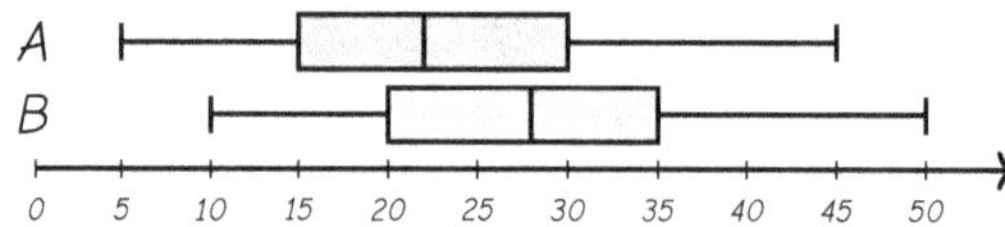

Which data set has a greater IQR?

   (A) *A (IQR = 15)*   (B) *B (IQR = 15)*

   (C) *They have the same IQR.*   (D) *Cannot be determined.*

29. *A bag has 5 orange balls, 3 purple balls, and 2 yellow balls. What is the probability of **not** drawing a yellow ball?*

   (A) $\dfrac{2}{10}$ (B) $\dfrac{5}{10}$

   (C) $\dfrac{8}{10}$ (D) $\dfrac{3}{10}$

30. *A circle graph shows pets owned by 120 students: Dogs 40%, Cats 30%, Fish 15%, Birds 15%. How many students own cats?*

   (A) 30   (B) 36

   (C) 40   (D) 48

 End of Practice Test 10 

Great job finishing the test!

☑ My Score

I got _____________ out of 30 questions right.

Check your answers in the Answer Key at the back of the book.

💡 Review any questions you missed. That's how we learn!

📊 Check Your Score Online!

Visit **ViewMath Academy** to enter your answers and see which topics you need to review. You can also explore lessons, take quizzes, track your scores, and save your progress!

viewmath.com/score/6.1.VA.25

Or go to **viewmath.com/score** and enter code: 6.1.VA.25

Answer Key & Explanations

Answer Key

First try each test on your own, then check your work here.

✅ Practice Test 1 — Answer Key

1 5   **2** C   **3** B   **4** A   **5** D   **6** 2.4   **7** C   **8** A   **9** $(-6, 2)$

10 8 blocks   **11** C   **12** 30 *square cm*   **13** $5(3n + 2)$ **14** B

15 6, 7, and 10 *(or any three numbers that are 6 or greater)*   **16** B   **17** $13.5\ m^2$ **18** 12 ft

19 B   **20** B   **21** B   **22** A   **23** C   **24** C   **25** C   **26** C   **27** B

28 Not always.   **29** B   **30** $450

💡 Time to Learn! 💡

*Review the explanations below, **especially for the questions you missed**.*

Understanding why each answer is correct builds stronger problem-solving skills.

Tip: *Circle any questions you got wrong, then read their explanation carefully.*

📖 Practice Test 1 — Detailed Explanations

1 *Each part* $= 20 \div 4 = 5$. *Vegetables* $= 1 \times 5 = 5$.

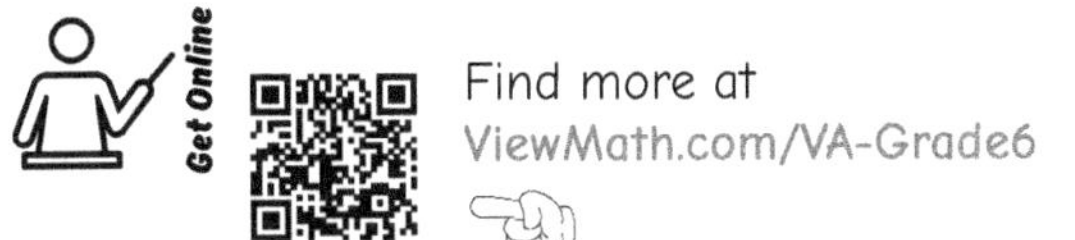

2. *From the graph, each hour earns \$10 (2 units on the y-axis = \$10). In 6 hours:* $6 \times 10 = \$60.$

3. *Ratio:* $1 : 4.$ *When* $x = 3$: $y = 3 \times 4 = 12.$

4. $0.15 \times 60 = 9.$

5. *The student flipped the first fraction* ($\frac{4}{5} \to \frac{5}{4}$) *instead of the second* ($\frac{2}{3} \to \frac{3}{2}$). *The correct answer is* $\frac{4}{5} \times \frac{3}{2} = \frac{12}{10} = \frac{6}{5}.$

6. *Move the decimal two places in both numbers:* $28.8 \div 12 = 2.4.$ *Check:* $2.4 \times 0.12 = 0.288.$

7. *On the number line, the first position that has both a circle and a triangle is* $12.$ *That means* 12 *is the smallest number that is a multiple of both* 3 *and* $4,$ *so the LCM is* $12.$

8. $8 \times 27 = 8(20 + 7) = 8 \times 20 + 8 \times 7 = 160 + 56 = 216.$

9. *Reflecting* $(6, -2)$ *across the y-axis flips the x-sign:* $(-6, -2).$ *Then reflecting across the x-axis flips the y-sign:* $(-6, 2).$ *A double reflection across both axes changes the signs of both coordinates:* $(x, y) \to (-x, -y).$

10. *Both locations share* $y = 4,$ *so the distance is* $|-6 - 2| = |-8| = 8$ *blocks.*

11. *The four boxes show* 4 *terms:* $7m, 2n, 4,$ *and* $3m.$

12. $A = \frac{1}{2} \times 10 \times 6 = \frac{1}{2} \times 60 = 30$ *square cm.*

13. *GCF of* 15 *and* 10 *is* $5.$ *Factor:* $5(3n + 2).$

14. *The one-time fee is \$25. Monthly cost is* $10m.$ *Total:* $25 + 10m.$

Find more at
ViewMath.com/VA-Grade6

15. Any number greater than or equal to 6 is a solution.

16. Open circle means -2 is NOT included ($<$ or $>$). Shading left means less than: $x < -2$.

17. $A = \frac{1}{2} \times 4.5 \times 6 = 13.5\ m^2$.

18. $96 = b \times 8$, so $b = 96 \div 8 = 12\ ft$.

19. Box A volume $= 6 \times 4 \times 5 = 120$. Box B: $120 = 10 \times 3 \times h = 30h$, so $h = 4$.

20. A reflection preserves the size and shape. The coordinates change, and the orientation flips, but the triangle stays congruent to the original.

21. Base $= |2 - (-4)| = 6$. Height $= |5 - 0| = 5$. Area $= \frac{1}{2} \times 6 \times 5 = 15$ square units.

22. $SA = 2(12)(8) + 2(12)(5) + 2(8)(5) = 192 + 120 + 80 = 392\ in^2$.

23. Across the y-axis: change the sign of x, keep y. $(2, -6) \to (-2, -6)$.

24. Radius $= 20 \div 2 = 10\ ft$. $A = \pi r^2 = 3.14 \times 10^2 = 3.14 \times 100 = 314\ ft^2$. Choice A is the circumference. Choice D uses πd^2.

25. Even number of values. The two middle values are 30 and 40. Median $= (30 + 40) \div 2 = 35$.

26. $MAD = (4 + 2 + 0 + 2 + 4) \div 5 = 12 \div 5 = 2.4$.

27. The 10–19 bar has height 10, the tallest bar.

28   *A single outlier can make the range very large while the rest of the data is tightly clustered. The IQR gives a better picture of overall spread. For example, $\{1, 50, 51, 52, 53\}$ has range 52 but most values are close together.*

29   *Numbers ≤ 3: $1, 2, 3$. That is 3 out of 6.* $P(\leq 3) = \dfrac{3}{6} = \dfrac{1}{2}$.

30 $15\% \times 3{,}000 = 0.15 \times 3{,}000 = 450$. *The family spends $450 on entertainment.*

✅ Practice Test 2 — Answer Key

1   C   **2**   *12, 16, and 30*   **3**   *Part A: Runner A $= 3 : 4$; Runner B $= 2 : 3$.*   *Part B: Runner A is faster.*

4   C   **5**   B   **6**   A   **7**   B   **8** $36 \div 6 = 6$, *not 4; the GCF is 18, so* $54 + 36 = 18(3 + 2)$

9   C

10   *8 units. Subtract the x-coordinates:* $5 - (-3) = 8$. *Take the absolute value:* $|8| = 8$. *The absolute value ensures dis*

11   D   **12**   25   **13** $6x + 12$

14   *Answers vary. Example: You buy n notebooks at $5 each and pay $3 for shipping.*   **15**   D   **16**   B

17   B   **18**   A   **19** $216\ in^3$ **20**   B   **21**   A   **22** $268\ m^2$ **23**   C   **24** $37.68   **25**   B

26   *Range $= 7$, IQR $= 2$*   **27**   C   **28**   A   **29**   C   **30**   C

💡 Time to Learn! 💡

*Review the explanations below, **especially for the questions you missed**.*

Understanding why each answer is correct builds stronger problem-solving skills.

Tip: *Circle any questions you got wrong, then read their explanation carefully.*

📖 Practice Test 2 — Detailed Explanations

1. *Girls have 6 parts $= 24$, so each part $= 24 \div 6 = 4$. Boys have 4 parts $= 4 \times 4 = 16$.*

2. *Row 2: $4 \times 2 = 8$, so $6 \times 2 = 12$. Row 3: $6 \times 4 = 24$, so $4 \times 4 = 16$. Row 4: $4 \times 5 = 20$, so $6 \times 5 = 30$.*

3. *Runner A: 3 laps in 4 min $= 0.75$ laps/min. Runner B: 2 laps in 3 min ≈ 0.667 laps/min. $0.75 > 0.667$, so Runner A is faster. On the graph, Runner A's line is steeper.*

4. *$0.25 \times ? = 20$. Divide: $20 \div 0.25 = 80$.*

5. *The correct answer is $\frac{3}{5} \times \frac{9}{2} = \frac{27}{10}$. The student computed $\frac{3}{5} \times \frac{2}{9} = \frac{6}{45}$, multiplying without flipping.*

6. *0.60×3.5: as whole numbers $60 \times 35 = 2{,}100$. Three decimal places: $2.100 = \$2.10$.*

7. *Multiples of 4: $4, 8, 12, \ldots$ Multiples of 6: $6, 12, \ldots$. The smallest number in both lists is 12.*

8. *Marcus divided $36 \div 6$ incorrectly, getting 4 instead of 6. Also, 6 is not the GCF. The GCF of 54 and 36 is 18. Divide: $54 \div 18 = 3$ and $36 \div 18 = 2$. Correct form: $18(3 + 2) = 90$.*

9. *The x-coordinate -3 tells you to move 3 units to the left, and the y-coordinate 4 tells you to move 4 units up. Choice D swaps the coordinates.*

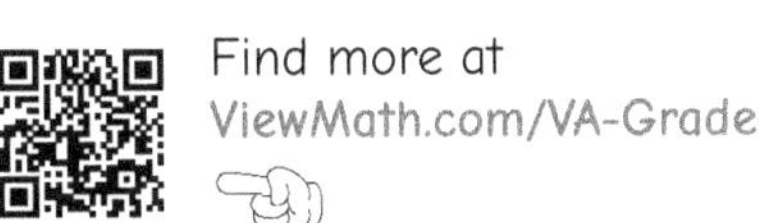

Find more at
ViewMath.com/VA-Grade6

10. Since both points share $y = -1$, the distance is $|5 - (-3)| = |5 + 3| = |8| = 8$. Absolute value removes any negative sign, because distance measures how far apart two points are and cannot be negative.

11. A constant has no variable. 5 is the only term without a variable.

12. $4^2 + 2(4) + 1 = 16 + 8 + 1 = 25$.

13. Perimeter $= 2(2x + 5) + 2(x + 1) = 4x + 10 + 2x + 2 = 6x + 12$.

14. $5n$ represents a cost per item and 3 represents a fixed fee.

15. "At least 48" means 48 or more: $h \geq 48$.

16. Both shade to the right. The only difference is the circle at 5: open for $>$, closed for $\geq$.

17. $A = \frac{1}{2} \times 12 \times 9 = 54$ ft^2.

18. $A = 6.5 \times 4 = 26$ ft^2.

19. $V = 6 \times 6 \times 6 = 216$ in^3.

20. Across the x-axis: change the sign of y. $(4, 7) \to (4, -7)$.

21. Length $= |6 - (-3)| = 9$. Width $= 54 \div 9 = 6$ units.

22. $SA = 2(10)(8) + 2(10)(3) + 2(8)(3) = 160 + 60 + 48 = 268$ m^2.

23. Right means add to x; down means subtract from y. The rule is $(x, y) \to (x + 4, y - 3)$.

Find more at
ViewMath.com/VA-Grade6

ViewMath.com

24 $Circumference = \pi d = 3.14 \times 6 = 18.84$ ft. $Cost = 18.84 \times 2 = \$37.68$.

25 $Mean = (70 + 80 + 90 + 85 + 75) \div 5 = 400 \div 5 = 80$.

26 Data: $3, 4, 4, 5, 5, 5, 6, 6, 6, 6, 7, 7, 8, 10$. 14 values. Range $= 10 - 3 = 7$. Median $= (6 + 6) \div 2 = 6$. Lower half: $3, 4, 4, 5, 5, 5, 6$, $Q1 = 5$. Upper half: $6, 6, 7, 7, 8, 10$... wait, let me recount. 14 values: lower 7: $3, 4, 4, 5, 5, 5, 6$ ☒ $Q1 = 5$. Upper 7: $6, 6, 6, 7, 7, 8, 10$ ☒ $Q3 = 7$. $IQR = 7 - 5 = 2$.

27 The mode is the value with the highest frequency — it appears more often than any other value.

28 Same medians, but Athlete A's range (1.2) is smaller — the times vary less, indicating more consistency.

29 Event Y is positioned closest to 0.75 on the probability scale.

20 All sectors must total 100%. $100 - 40 - 25 - 15 = 20\%$.

☑ Practice Test 3 — Answer Key

1 C **2** C **3** B **4** B **5** D **6** C **7** B **8** C **9** B **10** B

11 C **12** 16 **13** A **14** 7d; earnings for 8 dogs $= \$56$ **15** D **16** D **17** C

18 C **19** B **20** A **21** A **22** $108 \ cm^2$ **23** D **24** C **25** C **26** 6

27 C **28** B **29** C **30** C

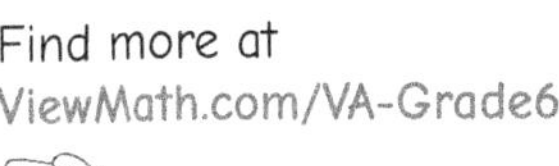
Find more at
ViewMath.com/VA-Grade6

💡 **Time to Learn!** 💡

*Review the explanations below, **especially for the questions you missed**.*

Understanding why each answer is correct builds stronger problem-solving skills.

Tip: *Circle any questions you got wrong, then read their explanation carefully.*

📖 Practice Test 3 — Detailed Explanations

1. *The question asks flour to eggs. Flour $= 3$, eggs $= 2$. The ratio is $3 : 2$.*

2. *$6 \times 3 = 18$ cups of flour, so $4 \times 3 = 12$ eggs.*

3. *The ratio is $6 : 2 = 3 : 1$. When $x = 15$: $y = 15 \div 3 = 5$.*

4. *Increase: $0.10 \times \$25 = \$2.50$. New price: $\$25 + \$2.50 = \$27.50$.*

5. *$\dfrac{4}{5} \div \dfrac{2}{5} = \dfrac{4}{5} \times \dfrac{5}{2} = \dfrac{20}{10} = 2$ servings.*

6. *Line up the decimals: $20.30 - 8.57$. Regroup as needed: $20.30 - 8.57 = 11.73$.*

7. *$36 \div 8 = 4.5$, so 8 does not divide evenly into 36. It is a factor of 24 but not of 36, so it is not a common factor. The other choices (4, 6, 12) all divide evenly into both 24 and 36.*

8. *The GCF of 24 and 18 is 6. Divide each term: $24 \div 6 = 4$ and $18 \div 6 = 3$. So $24 + 18 = 6(4 + 3)$. Choices A and B use common factors that are not the GCF. Choice D has the wrong quotient for $18 \div 6$.*

9. *When the y-coordinate is 0, the point lies on the x-axis. The point $(-6, 0)$ is 6 units to the left of the origin on the x-axis. A common mistake is thinking a negative x-value means the point is on the y-axis.*

Find more at
ViewMath.com/VA-Grade6

10   Choice A: $|3 - 7| = 4$. Choice B: $|-2 - 4| = 6$. Choice C: $|-3 - 2| = 5$. Choice D: $|-1 - 3| = 4$. The greatest distance is 6 in Choice B.

11 $5m - 2n + 7 + m$ has 4 terms: $5m$, $2n$, 7, and m.

12 $8^2 = 64$. Then $64 \div 4 = 16$.

13 $4(n + 2) = 4n + 8$ by the distributive property. They are equivalent for every value of n.

14 $7(8) = 56$ dollars.

15 $2.5 \geq 3$ is false. $2.5 < 3$, so it is not a solution.

16 $\leq$ includes -1 (closed circle). Less than -1 is to the left (shade left).

17   The triangle has base 6 and height 4. Area $= \frac{1}{2} \times 6 \times 4 = 12$ square units.

18   The height of a parallelogram is the perpendicular distance from the base to the opposite side, not the slanted side.

19   Total volume $= 5 \times 3 \times 2 = 30$ m^3. Half-full: $30 \div 2 = 15$ m^3.

20   Across the x-axis, change the sign of each y-coordinate: $(1, -3)$, $(5, -3)$, $(5, -7)$.

21   Rectangle area: $7 \times 3 = 21$. Triangle area: $\frac{1}{2} \times 2 \times 3 = 3$. Remaining: $21 - 3 = 18$ square units.

22   2 triangles: $2 \times \frac{1}{2}(3)(4) = 12$ cm^2. 3 rectangles: $8 \times 3 + 8 \times 4 + 8 \times 5 = 24 + 32 + 40 = 96$ cm^2. Total: $12 + 96 = 108$ cm^2.

23 A translation slides every point the same distance in the same direction. Size, shape, and orientation are all preserved.

24 The area of a circle is $A = \pi r^2$. Choices A and B are circumference formulas. Choice D incorrectly doubles the area formula.

25 Total $= 15 \times 6 = 90$. Remove 9: $90 - 9 = 81$. New mean $= 81 \div 5 = 16.2$.

26 Distances from 60: $10, 5, 0, 5, 10$. MAD $= (10 + 5 + 0 + 5 + 10) \div 5 = 30 \div 5 = 6$.

27 Every value is 5, so 5 is the value that appears most often. The mode is 5.

28 A small IQR means the middle 50% is tightly packed. A large range means the minimum and maximum are far apart, indicating extreme values stretch the data.

29 There are 6 equally likely outcomes. Only 1 of them is a 4. $P(4) = \dfrac{1}{6}$.

30 Savings $+$ Charity $= 30\% + 10\% = 40\%$. Then $40\% \times \$40 = 0.40 \times 40 = \$16$.

✔ Practice Test 4 — Answer Key

1 B	2 64	3 C	4 360	5 C	6 B	7 A	8 306	9 C	10 D

11 B	12 30	13 B	14 $n + 12 - 5$ or $n + 7$	15 C	16 C	17 B	18 A

19 B	20 D	21 B	22 C	23 B	24 200.96 m^2	25 B	26 C	27 B

28 A   29 A: 0 (impossible); B: $\dfrac{1}{2}$ (equally likely); C: $\dfrac{3}{4}$ (likely); D: 1 (certain)   30 B

Find more at
ViewMath.com/VA-Grade6

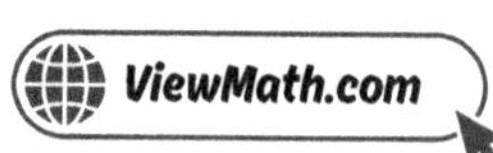

💡 **Time to Learn!** 💡

*Review the explanations below, **especially for the questions you missed**.*

Understanding why each answer is correct builds stronger problem-solving skills.

Tip: *Circle any questions you got wrong, then read their explanation carefully.*

📖 Practice Test 4 — Detailed Explanations

1. "3 pencils for every 1 eraser" means pencils to erasers is $3 : 1$.

2. $3 \times 8 = 24$ red, so white $= 5 \times 8 = 40$. Total $= 24 + 40 = 64$.

3. $5 : 15$ simplifies to $1 : 3$.

4. $0.72 \times 500 = 360$.

5. Keep, Change, Flip: $\dfrac{1}{2} \times \dfrac{4}{1} = \dfrac{4}{2} = 2$.

6. Multiply as whole numbers: $12 \times 3 = 36$. Count decimal places: 1.2 has 1 and 0.3 has 1, total 2. Place the decimal: 0.36.

7. Find the GCF of 72 and 96. Factors of 72: $1, 2, 3, 4, 6, 8, 9, 12, 18, 24, 36, 72$. Factors of 96: $1, 2, 3, 4, 6, 8, 12, 16, 24, 32, 48, 96$. Common factors include $1, 2, 3, 4, 6, 8, 12, 24$. The greatest is 24 inches.

8. Break 34 into $30 + 4$. Then $9 \times 34 = 9(30 + 4) = 9 \times 30 + 9 \times 4 = 270 + 36 = 306$.

9. Quadrant III contains points where both coordinates are negative $(-, -)$. Only $(-4, -7)$ has both coordinates negative. Choice A is in Quadrant II, B is in Quadrant IV, and D is in Quadrant I.

Find more at
ViewMath.com/VA-Grade6

ViewMath.com

10 The two points share the same x-coordinate, so the distance is the absolute difference of the y-coordinates: $|-3-5| = |-8| = 8$. Choice A forgets the absolute value. Choice B subtracts the x-coordinates. Choice C subtracts the wrong pair.

11 The expression $8(x+y) = 8 \times (x+y)$. There are 2 factors: 8 and $(x+y)$.

12 $A = \frac{1}{2} \times 12 \times 5 = 6 \times 5 = 30$.

13 The table confirms the same output for every input, and the distributive property proves equivalence: $2(x+3) = 2x + 6$.

14 Start with n, add 12, subtract 5: $n + 12 - 5 = n + 7$.

15 "Fewer than 20" means less than 20: $y < 20$.

16 $x \geq 3$ includes 3, so the circle should be closed (filled in), not open.

17 $P = \frac{1}{2} \times 10 \times 6 = 30$. $Q = \frac{1}{2} \times 10 \times 8 = 40$. Difference $= 40 - 30 = 10$ cm^2.

18 $11 \times 3 = 33$ cm^2. Maria got 16.5 because she divided by 2, using the triangle formula by mistake.

19 $V = l \times w \times h$. If h becomes $2h$, then $V_{new} = l \times w \times 2h = 2(lwh)$, so the volume doubles.

20 $(2,5)$ reflects to $(2,-5)$. Distance $= |5 - (-5)| = 10$ units.

21 Base $= |8-2| = 6$. Height $= |7-1| = 6$. Area $= \frac{1}{2} \times 6 \times 6 = 18$ square units.

22 2 triangular bases $+3$ rectangular faces $= 5$ faces total.

23   Right 4, down 2: add 4 to x, subtract 2 from y. $C(-3, 6) \rightarrow C'(-3 + 4, 6 - 2) = (1, 4)$.

24 $A = \pi r^2 = 3.14 \times 8^2 = 3.14 \times 64 = 200.96 \ m^2$.

25   With an even number of values, the median is the average of the two middle values: $(18 + 20) \div 2 = 19$.

26   IQR measures the spread of the middle 50%. A smaller IQR (A) means those values are closer together.

27   The size of the data set and whether it is numerical or categorical determine the best display. Categorical ⊠ frequency table/bar graph. Small numerical ⊠ dot plot. Large numerical ⊠ histogram.

28 $Q_1 = (15 + 18)/2 = 16.5$. $Q_3 = (25 + 28)/2 = 26.5$. $IQR = 26.5 - 16.5 = 10$.

29   Event A: A standard die has faces 1–6, so rolling a 7 is impossible; $P = 0$. Event B: $P(tails) = \dfrac{1}{2}$. Event C: $P(red) = \dfrac{6}{8} = \dfrac{3}{4} = 0.75$. Event D: All faces (1–6) are less than 7, so $P = 1$ (certain).

30 $100\% - 28\% - 34\% = 38\%$. Candidate C received 38% of the votes.

☑ Practice Test 5 — Answer Key

1 10   **2** B   **3** B   **4** D   **5** A   **6** $16.43   **7** A   **8** B   **9** A

10 $y = 13$ or $y = -3$ **11** 6a, 4b, and 11   **12** B   **13** $11m + 2$ **14** B   **15** C

16 $x = 3$ IS a solution to $x \leq 3$ because $3 \leq 3$ is true. $x = 3$ is NOT a solution to $x < 3$ because $3 < 3$ is false.

17 B   **18** B   **19** B   **20** D   **21** 64 square units   **22** B   **23** A   **24** B   **25** B

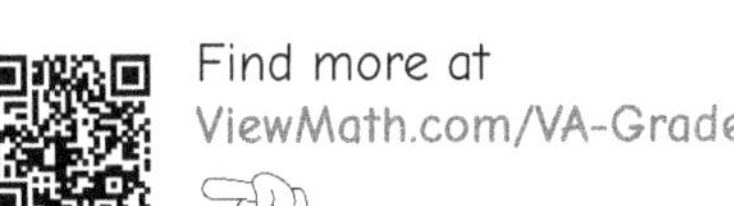
Find more at
ViewMath.com/VA-Grade6

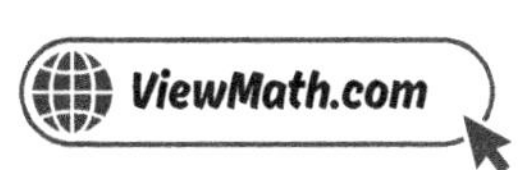

 B    Total = 30, Mode = Pizza    B    (a) $\frac{3}{8}$ (b) $\frac{1}{4}$ (c) $\frac{5}{8}$ **30** C

💡 Time to Learn! 💡

*Review the explanations below, **especially for the questions you missed**.*

Understanding why each answer is correct builds stronger problem-solving skills.

***Tip:** Circle any questions you got wrong, then read their explanation carefully.*

📖 Practice Test 5 — Detailed Explanations

1 Total parts $= 3 + 2 = 5$. Each part $= 25 \div 5 = 5$. Swimming $= 2 \times 5 = 10$.

2 $2 : 3$ multiplied by 3 gives $6 : 9$. The other pairs do not simplify to $2 : 3$.

3 Ratio $= 3 : 2$. Double: $(6, 4)$. Triple: $(9, 6)$, not $(9, 4)$ or $(9, 8)$.

4 Backpack: $0.15 \times \$40 = \$6$. Shoes: $0.10 \times \$60 = \$6$. Hat: $0.25 \times \$20 = \$5$. Backpack and shoes are tied at $\$6$.

5 $\dfrac{5}{6} \times \dfrac{3}{1} = \dfrac{15}{6} = \dfrac{5}{2} = 2\dfrac{1}{2}$.

6 $12.95 + 3.48 = 16.43$. Hundredths: $5 + 8 = 13$, write 3, carry 1. Tenths: $9 + 4 + 1 = 14$, write 4, carry 1. Ones: $2 + 3 + 1 = 6$. Tens: 1.

7 Factors of 36: $1, 2, 3, 4, 6, 9, 12, 18, 36$. Factors of 48: $1, 2, 3, 4, 6, 8, 12, 16, 24, 48$. Common factors: $1, 2, 3, 4, 6, 12$. The greatest is 12.

8   *Total* $= 24+16$. *The GCF of 24 and 16 is 8. Divide:* $24 \div 8 = 3$ *and* $16 \div 8 = 2$. *So* $24+16 = 8(3+2) = 8 \times 5 = \$40$. *Choice A uses 4 instead of the GCF.*

9   *Reflecting across the y-axis changes the sign of the x-coordinate only. So $(-5, 2)$ becomes $(5, 2)$. Choice B reflects across the x-axis. Choice C reflects across both axes.*

10   *The points share $x = -2$, so the distance is $|y - 5| = 8$. This means $y - 5 = 8$ or $y - 5 = -8$. Solving: $y = 13$ or $y = -3$. Both values give a distance of 8.*

11   *Terms are separated by $+$ or $-$ signs: $6a$, $4b$, and 11.*

12 $5(7 - 3) = 5(4) = 20$.

13   *Distribute: $8m + 2 + 3m$. Combine: $8m + 3m = 11m$. Result: $11m + 2$.*

14   *Start with 200 and lose 5 each day: $200 - 5d$.*

15   *$5 > 5$ is false. 5 equals 5, but is not greater than 5. It would be a solution to $x \geq 5$.*

16   *$\leq$ includes the boundary value; $<$ does not.*

17   *Rectangle area $= 8 \times 6 = 48$ in^2. Each triangle is half: $48 \div 2 = 24$ in^2.*

18   *The area of a parallelogram is $A = b \times h$, the same as a rectangle.*

19 $V = 3 \times 2 \times 2 = 12$ ft^3.

20 $|4 - (-2)| = |6| = 6$ *units.*

21 Bottom rectangle: $10 \times 4 = 40$. Top-left rectangle: $6 \times (8 - 4) = 6 \times 4 = 24$. Total: $40 + 24 = 64$ square units.

22 $SA = 2(10)(3) + 2(10)(7) + 2(3)(7) = 60 + 140 + 42 = 242 \ m^2$.

23 Right 7: $-5 + 7 = 2$. Up 4: $-2 + 4 = 2$. So $K'(2, 2)$.

24 Pi (π) is defined as the ratio of a circle's circumference to its diameter. For every circle, $\pi = \dfrac{C}{d} \approx 3.14$.

25 The mean uses every value in its calculation, so one extreme value changes it significantly. The median depends only on the middle position.

26 Distances from 24: $4, 2, 0, 2, 4$. $MAD = (4 + 2 + 0 + 2 + 4) \div 5 = 12 \div 5 = 2.4$.

27 $12 + 8 + 4 + 6 = 30$. Pizza has the highest frequency (12), so it is the mode.

28 Same median (same typical sales). Store A's range ($200) is much smaller than Store B's ($800), making Store A more consistent.

29 There are 8 equal sections: 3 red, 2 blue, 3 green. (a) $P(red) = \dfrac{3}{8}$. (b) $P(blue) = \dfrac{2}{8} = \dfrac{1}{4}$. (c) $P(not\ green) = 1 - \dfrac{3}{8} = \dfrac{5}{8}$.

30 $40\% - 25\% = 15\%$. Then $15\% \times 600 = 0.15 \times 600 = 90$ more students in Grade 7.

✅ Practice Test 6 — Answer Key

1 Bananas to yogurt: $3 : 4$; Yogurt to bananas: $4 : 3$  B  A  B  D  C

Find more at
ViewMath.com/VA-Grade6

 D    B    C    C    D    C

 The student forgot to distribute 4 to the 3. Correct: $8x + 12$  B    C    B    $54\ in^2$

 $9\ ft^2$  B    $(0, -4)$ and $(0, 3)$ B $348\ in^2$ C   B   D

 A   B   B   C   C

💡 *Time to Learn!* 💡

*Review the explanations below, **especially for the questions you missed**.*

Understanding why each answer is correct builds stronger problem-solving skills.

Tip: *Circle any questions you got wrong, then read their explanation carefully.*

📖 *Practice Test 6 — Detailed Explanations*

1. *"3 bananas for every 4 cups of yogurt" gives $3 : 4$. Flip the order for yogurt to bananas: $4 : 3$.*

2. *$3 \times 2 = 6$, so $5 \times 2 = 10$.*

3. *The ratio is $2 : 3$. Choice A continues with $6 : 9 = 2 : 3$. Choice B has $6 : 8$ which is not $2 : 3$.*

4. *10% means one-tenth. $350 \div 10 = 35$.*

5. *$\dfrac{7}{8} \div \dfrac{1}{4} = \dfrac{7}{8} \times \dfrac{4}{1} = \dfrac{28}{8} = \dfrac{7}{2} = 3\dfrac{1}{2}$ laps.*

6. *Move the decimal one place in both numbers: $45.6 \div 6 = 7.6$. Check: $7.6 \times 0.6 = 4.56$.*

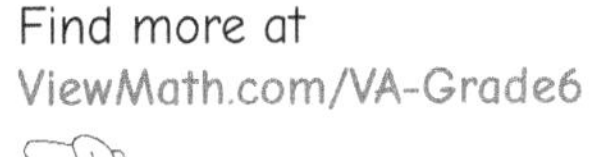

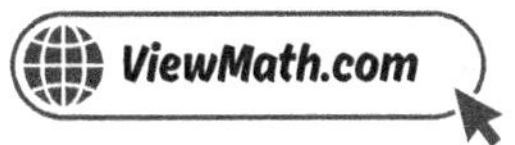

7 Find the LCM of 8 and 10. Multiples of 8: $8, 16, 24, 32, 40, \ldots$ Multiples of 10: $10, 20, 30, 40, \ldots$ LCM $= 40$. Both buses arrive together again after 40 minutes.

8 The GCF of 12 and 28 is 4. Divide: $12 \div 4 = 3$ and $28 \div 4 = 7$. So $12 + 28 = 4(3 + 7)$. Choice C forgets to divide 28 by the GCF.

9 Quadrant III has sign convention $(-, -)$. Since both $-3 < 0$ and $-6 < 0$, the point $(-3, -6)$ is in Quadrant III.

10 The points share $x = 1$, so the distance is $|-2 - 5| = |-7| = 7$ units. Choice A subtracts the y-values incorrectly as $5 - 2$. Choice B forgets the absolute value. Choice D adds the absolute values $|-2| + |5| + |1| + |1|$.

11 $x = 1 \cdot x$, so the coefficient is 1. The expression has 2 terms: x and 5.

12 When $x = 3$: $2(3) + 3 = 9$, not 8. The output for $x = 3$ is incorrect.

13 $4(2x + 3) = 4 \times 2x + 4 \times 3 = 8x + 12$, not $8x + 3$.

14 $12p$ means 12 times each person. So 12 is the cost per person, and 6 is a flat fee.

15 $n \leq 8$ means n is 8 or less. $8 \leq 8$ is true. 8.5, 9, and 10 are all greater than 8.

16 $\leq$ means 5 is included (closed circle). Less than 5 is to the left (shade left).

17 For a right triangle the legs are the base and height. $A = \frac{1}{2} \times 9 \times 12 = 54$ in^2.

18 $A = \frac{1}{2}(2 + 4)(3) = \frac{1}{2}(6)(3) = 9$ ft^2.

19 $2\frac{1}{2} = 2.5$. $V = 2.5 \times 4 \times 3 = 30$ ft^3.

Find more at
ViewMath.com/VA-Grade6

20 Both points are on the y-axis ($x = 0$). Reflecting across the y-axis: $(0, -4) \to (0, -4)$ and $(0, 3) \to (0, 3)$. Points on the axis of reflection don't move.

21 Width $= 56 \div 8 = 7$ units.

22 $SA = 2(15)(6) + 2(15)(4) + 2(6)(4) = 180 + 120 + 48 = 348$ in^2.

23 The y-coordinate stayed the same. The x-coordinate changed from -2 to 4: $4 - (-2) = 6$ units right. Note: reflection across the y-axis would give $(2, 7)$, not $(4, 7)$.

24 First find the radius: $r = 12 \div 2 = 6$ m. Then $A = \pi r^2 = 3.14 \times 6^2 = 3.14 \times 36 = 113.04$ m^2. Choice D uses the diameter instead of the radius in πd^2.

25 Mean $=$ sum $\div 5 = 20$, so sum $= 20 \times 5 = 100$.

26 Team A range: $68 - 60 = 8$. Team B range: $88 - 40 = 48$. Team A is much more consistent.

27 Histogram bins are consecutive numerical intervals with no values left out, so bars sit directly next to each other.

28 Week 2 median ($9,500$) is higher. Week 2 IQR ($1,200$) is smaller, meaning more consistent daily step counts.

29 A fair coin has 2 equally likely outcomes (heads or tails). $P(\text{heads}) = \dfrac{1}{2}$.

30 $40\% \times 200 = 0.40 \times 200 = 80$. So 80 students take the bus.

☑ ## Practice Test 7 — Answer Key

Find more at
ViewMath.com/VA-Grade6

 6    B    $(2,5)$ and $(8,20)$ *(or any equivalent pair)*    B    5    B

7 *36 days*    B    A    C

11 $4x + y + 10$ *(or any valid expression with the required features)*    B    D    C

15 $a \geq 16$ **16** B   **17** B   **18** B   **19** D   **20** $(-2,4)$ **21** B   **22** B   **23** $(8,3)$

24 C   **25** 26   **26** B   **27** *30 students. Modal interval: 70–79. Fraction* ≥ 70: $\frac{23}{30}$.   **28** B

29 B   **30** C

💡 Time to Learn! 

*Review the explanations below, **especially for the questions you missed**.*

Understanding why each answer is correct builds stronger problem-solving skills.

Tip: *Circle any questions you got wrong, then read their explanation carefully.*

📖 Practice Test 7 — Detailed Explanations

$21 \div 7 = 3$, *so multiply both by 3: iced tea* $= 2 \times 3 = 6$.

$2 \times 4 = 8$ *and* $3 \times 4 = 12$, *so* $8 : 12$ *is equivalent to* $2 : 3$.

Ratio $= 4 : 10 = 2 : 5$. *Half gives* $(2,5)$*; double gives* $(8,20)$.

$0.15 \times 4{,}000 = 600$ *children.*

$\dfrac{5}{6} \div \dfrac{1}{6} = \dfrac{5}{6} \times \dfrac{6}{1} = \dfrac{30}{6} = 5$ *birdhouses.*

Find more at
ViewMath.com/VA-Grade6

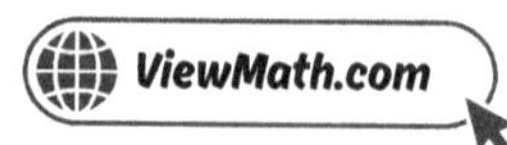

6. Line up the decimals: $4.36 + 9.70 = 14.06$. Fill the empty hundredths place with a zero before adding.

7. Find the LCM of 9 and 12. Multiples of 9: $9, 18, 27, 36, \ldots$ Multiples of 12: $12, 24, 36, \ldots$ $LCM = 36$. They will both be at the library again in 36 days.

8. $28 \div 7 = 4$ is correct, but $42 \div 7 = 6$, not 8. The correct form using 7 is $7(4+6) = 70$. Note: the GCF of 28 and 42 is actually 14, giving $14(2+3) = 70$.

9. In Quadrant I both coordinates are positive $(+,+)$. Since $5 > 0$ and $3 > 0$, the point $(5,3)$ is in Quadrant I.

10. The lights share $y = 5$, so the distance is $|-3 - 4| = |-7| = 7$ blocks. Choice B uses only the y-coordinate. Choice D adds $|-3| + |4| + 2$ or a similar error.

11. Answers vary. A sample: $4x + y + 10$ has 3 terms, one with coefficient 4, and constant 10.

12. $2^3 = 2 \times 2 \times 2 = 8$.

13. $3x + 2x = 5x$, not $5x^2$. Adding like terms keeps the variable as x, not x^2.

14. $50 + 40(3) = 50 + 120 = 170$ dollars.

15. "No fewer than 16" means 16 or more: $a \geq 16$.

16. $\leq$ means less than or equal to 6. Numbers less than 6 are to the left, so shade left.

17. $A = \frac{1}{2} \times 10 \times 6 = 30$ cm^2.

18. $A = \frac{1}{2}(9 + 15)(6) = \frac{1}{2}(24)(6) = 72$ in^2.

Find more at
ViewMath.com/VA-Grade6

19 $V = 20 \times 10 \times 12 = 2,400\ in^3$.

20 Across the y-axis: change the sign of x. $(2,4) \to (-2,4)$.

21 Bottom rectangle: $8 \times 3 = 24$. Left rectangle above: $4 \times (7-3) = 4 \times 4 = 16$. Total: 40. This matches option B.

22 $SA = 2(4)(4) + 2(4)(9) + 2(4)(9) = 32 + 72 + 72 = 176\ cm^2$.

23 Across the x-axis: keep x, change the sign of y. $(8,-3) \to (8,3)$.

24 Area of $P = \pi \times 4^2 = 16\pi$. Area of $Q = \pi \times 8^2 = 64\pi$. Ratio $= 64\pi \div 16\pi = 4$. When the radius doubles, the area quadruples because area depends on r^2.

25 Even number of values. Middle values are 24 and 28. Median $= (24 + 28) \div 2 = 26$.

26 MAD (Mean Absolute Deviation) is calculated by finding the average of the distances of each data value from the mean.

27 Total $= 2 + 5 + 10 + 8 + 5 = 30$. The 70–79 bar is tallest (10). Students ≥ 70: $10 + 8 + 5 = 23$. Fraction $= 23/30$.

28 The median describes a typical value (center) and the IQR describes how spread out the middle 50% of values are (spread). Together they summarize the data well.

29 $\frac{2}{5} = 0.4$ and $\frac{3}{4} = 0.75$. Since $0.75 > 0.4$, Event B is more likely.

30 $75\% = \frac{75}{100} = \frac{3}{4}$ after dividing the numerator and denominator by 25.

Find more at
ViewMath.com/VA-Grade6

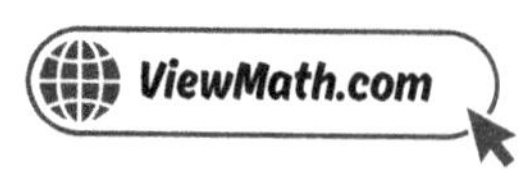

✅ Practice Test 8 — Answer Key

1 B   **2** Part A: 2 : 1;   Part B: 4 cups of sugar

3 False. Equivalent ratios always form a straight line through the origin.   **4** C   **5** $\frac{3}{2}$ or $1\frac{1}{2}$ **6** C

7 D   **8** C   **9** A   **10** Part A: 7 units;   Part B: 7 units;   Part C: 14 units   **11** B   **12** C

13 $6a + 10$ **14** $50 + 15w$; after 10 weeks: $200

15 Answers vary. Example: A classroom has at most 15 computers.   **16** $x < 12$ **17** 9 m   **18** D

19 60 cm^3 **20** C   **21** 24 square units   **22** C   **23** C   **24** 254.34 in^2

25 Mean ≈ 13.3, Median $= 6$. The median is better.   **26** C   **27** Mode $= 5000$, Range $= 4000$ **28** C

29 20%   **30** B

💡 Time to Learn! 💡

Review the explanations below, **especially for the questions you missed**.

Understanding why each answer is correct builds stronger problem-solving skills.

Tip: Circle any questions you got wrong, then read their explanation carefully.

📖 Practice Test 8 — Detailed Explanations

1 Dogs have 4 parts. Each part = 2 animals. Dogs = $4 \times 2 = 8$.

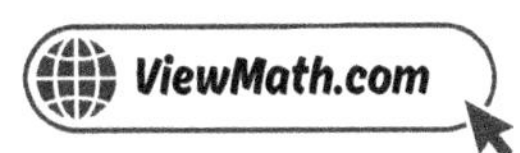

2   *Part A: From the graph, $(2,1)$ shows 2 cups of flour for every 1 cup of sugar. Part B: Following the pattern, $(8,4)$, so 4 cups of sugar.*

3   *When you multiply both parts of a ratio by the same number, the points increase at a constant rate, which produces a straight line through $(0,0)$.*

4 $\dfrac{12}{30} = \dfrac{40}{100} = 40\%.$

5 $\dfrac{9}{10} \times \dfrac{5}{3} = \dfrac{45}{30} = \dfrac{3}{2} = 1\dfrac{1}{2}.$

6   *1.4 has 1 decimal place and 0.3 has 1 decimal place, so the product needs $1 + 1 = 2$ decimal places. $14 \times 3 = 42 \to 0.42$. The student counted only 1 decimal place instead of 2.*

7   *Factors of 54: $1, 2, 3, 6, 9, 18, 27, 54$. Factors of 90: $1, 2, 3, 5, 6, 9, 10, 15, 18, 30, 45, 90$. Common factors: $1, 2, 3, 6, 9, 18$. The greatest is 18.*

8   *Total cost $= 7 \times 3 + 7 \times 2 = 7(3 + 2) = 7 \times 5 = \$35$. The common factor 7 is factored out.*

9   *Reflecting across the y-axis changes the sign of the x-coordinate while keeping the y-coordinate the same. So $(3,5)$ becomes $(-3,5)$. Choice B reflects across the x-axis instead.*

10   *Part A: Same $y = 3$, distance $= |-5 - 2| = |-7| = 7$. Part B: Same $x = 2$, distance $= |3 - (-4)| = |7| = 7$. Part C: Total $= 7 + 7 = 14$ units.*

11   *In $r - 5$, the coefficient of r is 1 (since $r = 1 \cdot r$), not 0. Row 2 is incorrect.*

12 $15 \div 3 + 8 = 5 + 8 = 13.$

13 $3a + 5a - 2a = 6a$ *and* $7 + 3 = 10$. *Result: $6a + 10$.*

Find more at
ViewMath.com/VA-Grade6

14 Each week adds 15: pattern is $50 + 15w$. When $w = 10$: $50 + 15(10) = 50 + 150 = 200$.

15 Any situation where a quantity is 15 or fewer works.

16 Open circle means the number is not included ($<$ or $>$). Shading left means less than: $x < 12$.

17 $45 = \frac{1}{2} \times 10 \times h$, so $45 = 5h$, giving $h = 9$ m.

18 $A = \frac{1}{2}(40 + 60)(30) = \frac{1}{2}(100)(30) = 1{,}500 \ m^2$.

19 $V = 2.5 \times 4 \times 6 = 60 \ cm^3$.

20 $A(-5, 4)$ reflected across the x-axis gives $A'(-5, -4)$. Negative x and negative y is Quadrant III.

21 Base $= |5 - (-3)| = 8$. Height $= |4 - (-2)| = 6$. Area $= \frac{1}{2} \times 8 \times 6 = 24$ square units.

22 A cube has 6 equal faces. $SA = 6 \times 5^2 = 6 \times 25 = 150 \ in^2$.

23 Across the y-axis: change the sign of x, keep y. $(-4, 2) \rightarrow (4, 2)$.

24 First find the radius: $r = 18 \div 2 = 9$ in. Then $A = \pi r^2 = 3.14 \times 9^2 = 3.14 \times 81 = 254.34 \ in^2$.

25 Mean: $(3 + 4 + 5 + 6 + 7 + 8 + 60) \div 7 = 93 \div 7 \approx 13.3$. Median: 6. The outlier 60 inflates the mean. The median (6) is more typical.

26 The range uses only the maximum and minimum values, so one extreme outlier can make the range very large. The IQR ignores the most extreme values.

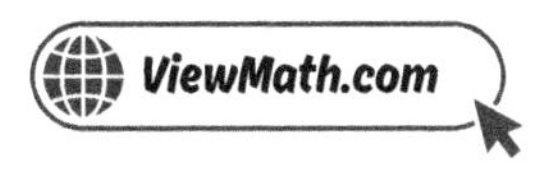

27 5000 has frequency 6 (highest). Range $= 7000 - 3000 = 4000$.

28 The median and IQR are resistant to outliers. The mean and range can be greatly affected by extreme values.

29 Total marbles: $4 + 6 + 10 = 20$. $P(green) = \dfrac{4}{20} = \dfrac{1}{5} = 0.20 = 20\%$.

30 Let T be the total. $15\% \times T = 450$, so $T = 450 \div 0.15 = 3{,}000$. The total donations were $\$3{,}000$.

✅ Practice Test 9 — Answer Key

1 C   2 C   3 A   4 B   5 A   6 B   7 C

8 Part A: 18;   Part B: $54 + 36 = 18(3 + 2)$;   Part C: 18 bags with 3 chocolate and 2 vanilla each   9 D

10 Part A: 9 units;   Part B: 5 units;   Part C: 14 units   11 B   12 A   13 B   14 C   15 C

16 C   17 C   18 B   19 A   20 B   21 B   22 B   23 B   24 B   25 B

26 C   27 B   28 B   29 B   30 75

💡 Time to Learn! 💡

Review the explanations below, **especially for the questions you missed**.

Understanding why each answer is correct builds stronger problem-solving skills.

Tip: Circle any questions you got wrong, then read their explanation carefully.

📖 Practice Test 9 — Detailed Explanations

Find more at
ViewMath.com/VA-Grade6

1. Total parts $= 1 + 3 + 2 = 6$. Each part $= 18 \div 6 = 3$. Blue $= 3 \times 3 = 9$.

2. $1 \times 3 = 3$ red, so $4 \times 3 = 12$ blue. Multiply both parts by 3.

3. Line A rises 4 for every 1 it moves right; Line B rises 3. A higher rise means steeper.

4. 25% of $80 = 0.25 \times 80 = \$20$.

5. $\dfrac{4}{9} \times \dfrac{3}{2} = \dfrac{12}{18} = \dfrac{2}{3}$.

6. Line up the decimals: $3.49 + 1.75 = 5.24$. Hundredths: $9 + 5 = 14$, write 4 carry 1. Tenths: $4 + 7 + 1 = 12$, write 2 carry 1. Ones: $3 + 1 + 1 = 5$.

7. Factors of 16: $1, 2, 4, 8, 16$. Factors of 40: $1, 2, 4, 5, 8, 10, 20, 40$. Common factors: $1, 2, 4, 8$. The greatest is 8.

8. Part A: Factors of 54: $1, 2, 3, 6, 9, 18, 27, 54$. Factors of 36: $1, 2, 3, 4, 6, 9, 12, 18, 36$. $GCF = 18$. Part B: $54 \div 18 = 3$ and $36 \div 18 = 2$, so $54 + 36 = 18(3 + 2) = 90$. Part C: She makes 18 bags, each with 3 chocolate cookies and 2 vanilla cookies.

9. The origin is the point where the x-axis and y-axis cross. Its coordinates are $(0, 0)$.

10. Part A: $A(-4, 3)$ and $B(5, 3)$ share $y = 3$. Distance $= |-4 - 5| = |-9| = 9$. Part B: $B(5, 3)$ and $C(5, -2)$ share $x = 5$. Distance $= |3 - (-2)| = |5| = 5$. Part C: Total $= 9 + 5 = 14$ units.

11. The term d means $1 \cdot d$, so the coefficient is 1. (The subtraction sign belongs to the operation, not the coefficient of d itself in the original expression.)

12. $3(4) + 7 = 12 + 7 = 19$.

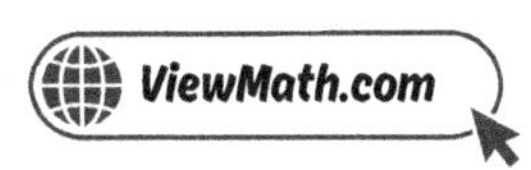

13. $10a - 2a = 8a$. The constant stays: $8a + 7$.

14. Area $=$ length $\times$ width $= l \times 5 = 5l$.

15. $t \leq 30$ means 30 or less, which is "no more than 30 degrees."

16. Open circle at 0 means 0 is not included. Shading right means greater than: $x > 0$.

17. $A = \frac{1}{2} \times 3.5 \times 4 = 7$ cm^2.

18. Area uses the height, not the slant side. $A = 12 \times 6 = 72$ m^2.

19. $V = \frac{1}{2} \times \frac{1}{2} \times \frac{1}{2} = \frac{1}{8}$ m^3.

20. Across the y-axis: only the x-coordinate changes sign. $(3, -8) \to (-3, -8)$. Sara also changed the y.

21. Base $= |5 - 1| = 4$. Height $= |9 - 1| = 8$. Area $= \frac{1}{2} \times 4 \times 8 = 16$ square units.

22. $SA = 2(6)(4) + 2(6)(3) + 2(4)(3) = 48 + 36 + 24 = 108$ cm^2.

23. The x-coordinate stayed the same. The y-coordinate changed from -1 to 5: $5 - (-1) = 6$ units up.

24. $C = \pi d = 3.14 \times 10 = 31.4$ cm. Choice A uses πr incorrectly as the full circumference, and choice C doubles the correct answer.

25. There are 15 data points. The median is the 8th value. Counting from 0: 2 zeros, 3 ones (5 so far), then 4 twos (9 so far). The 8th value is 2.

26   Adding 5 to every value shifts the max and min by the same amount, so the difference (range) stays unchanged.

27   The 25–49 bar has height 8, the tallest bar. It is the modal interval.

28   The range uses only the maximum and minimum. Changing any extreme value changes the range. However, the IQR only uses Q_1 and Q_3 and can stay the same even if extremes change. Still, the range is most directly influenced by outliers since it depends on the two most extreme values.

29 $P(cherry) = \dfrac{4}{10}$ and $P(grape) = \dfrac{4}{10}$. The probabilities are equal, so these events are equally likely.

30 $Swimming = 100\% - 40\% - 35\% = 25\%$. Then $25\% \times 300 = 0.25 \times 300 = 75$ students.

☑ Practice Test 10 — Answer Key

1 C   **2** B   **3** C   **4** C   **5** A   **6** Part A: 27.54 km;   Part B: 9.18 km per day

7 35   **8** $6(2 + 3 + 5)$ **9** D   **10** C   **11** (a) 9 and 1; (b) 5; (c) 4 terms   **12** A

13 Error in Step 2. Should be $6x + 12 + x$. Correct answer: $7x + 12$ **14** B   **15** C

16 Open circle at 1, shade to the right   **17** C   **18** 72 square units   **19** C   **20** C   **21** C

22 A   **23** $(3, 3)$ **24** B   **25** B   **26** Data set B   **27** B   **28** C   **29** C   **30** B

💡 Time to Learn! 💡

*Review the explanations below, **especially for the questions you missed**.*

Understanding why each answer is correct builds stronger problem-solving skills.

Tip: *Circle any questions you got wrong, then read their explanation carefully.*

📖 Practice Test 10 — Detailed Explanations

1. Apple has 3 parts, orange has 5 parts. Total parts = 8. Total ounces = $8 \times 4 = 32$.

2. The ratio is $2 : 50$. For 6 bags: $50 \times 3 = 150$ sq ft.

3. At \$2 per bottle, 4 bottles cost \$8, not \$10. The point $(4, 10)$ does not fit the ratio.

4. $\dfrac{45}{90} = \dfrac{1}{2} = 50\%$.

5. Dividing by $\dfrac{c}{d}$ is the same as multiplying by its reciprocal, $\dfrac{d}{c}$.

6. Part A: $8.76 + 12.30 + 6.48 = 27.54$ km. Part B: $27.54 \div 3 = 9.18$ km per day. Check: $9.18 \times 3 = 27.54$.

7. Multiples of 5: $5, 10, 15, 20, 25, 30, 35, \ldots$ Multiples of 7: $7, 14, 21, 28, 35, \ldots$ The smallest number in both lists is 35. Since 5 and 7 share no common factors, the LCM is $5 \times 7 = 35$.

8. The GCF of 12, 18, and 30 is 6. Divide: $12 \div 6 = 2$, $18 \div 6 = 3$, $30 \div 6 = 5$. So $12 + 18 + 30 = 6(2 + 3 + 5) = 6 \times 10 = 60$.

9. Quadrant IV has sign convention $(+, -)$. Since $2 > 0$ and $-5 < 0$, the point $(2, -5)$ is in Quadrant IV. A common error is confusing Quadrant IV with Quadrant II which has signs $(-, +)$.

Find more at
ViewMath.com/VA-Grade6

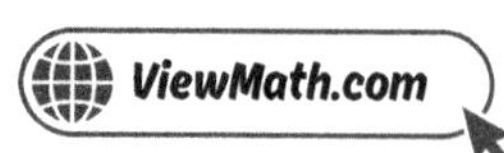

10 The points share the same x-coordinate, so the distance is $|-6-(-1)| = |-6+1| = |-5| = 5$. Choice B forgets the absolute value. Choice D adds the absolute values $6+1$ instead of subtracting.

11 Terms with y: $9y$ (coefficient 9) and y (coefficient 1). The constant is 5. There are 4 terms total.

12 $2(3(5)-1) = 2(15-1) = 2(14) = 28$.

13 The student distributed 3 to $2x$ but not to 4. Correct: $3 \times 4 = 12$. So $6x + 12 + x = 7x + 12$.

14 In 4 days you read $4r$ pages. Remaining: $100 - 4r$.

15 $0 > -2$ is true, so $x = 0$ is a solution. -2 is not greater than -2, and -3 and -2.5 are less than -2.

16 $>$ means 1 is not included (open circle). Greater values are to the right.

17 $A = \frac{1}{2} \times 2.4 \times 5 = 6 \ m^2$.

18 Rectangle: $10 \times 4 = 40$. Trapezoid: $\frac{1}{2}(10+6)(4) = \frac{1}{2}(16)(4) = 32$. Total: $40 + 32 = 72$ square units.

19 Each 1-inch cube takes up 1 in^3, so the number of cubes equals the volume: $12 \times 8 \times 6 = 576$.

20 Across the y-axis: change the sign of x. $(-3, 2) \to (3, 2)$.

21 Length $= 6$, width $= 4$. Area $= 6 \times 4 = 24$ square units.

22 A: $2(6) + 2(20) + 2(30) = 112$. B: $6(16) = 96$. C: $2(5) + 2(8) + 2(40) = 106$. D: $2(9) + 2(18) + 2(18) = 90$. Option A is greatest.

23 Reflect across the y-axis: $(-3, 5) \rightarrow (3, 5)$. Translate 2 units down: $(3, 5 - 2) = (3, 3)$.

24 Radius $= 10 \div 2 = 5$ cm. Full circle area $= \pi r^2 = 3.14 \times 25 = 78.5$ cm^2. Semicircle area $= 78.5 \div 2 = 39.25$ cm^2. Choice C is the full circle area.

25 Original mean: $(58 + 60 + 62 + 64) \div 4 = 61$. New mean: $(58 + 60 + 62 + 64 + 82) \div 5 = 326 \div 5 = 65.2$.

26 A larger MAD means values are farther from the mean on average. Data set B's MAD of 10 indicates more variability than A's MAD of 2.

27 7 appears 2 times, 8 appears 4 times, 9 appears 3 times, 10 appears 1 time. The mode is 8.

28 A: IQR $= 30 - 15 = 15$. B: IQR $= 35 - 20 = 15$. Both have IQR $= 15$.

29 Total balls: $5 + 3 + 2 = 10$. $P(\text{yellow}) = \dfrac{2}{10}$. $P(\text{not yellow}) = 1 - \dfrac{2}{10} = \dfrac{8}{10}$.

30 $30\% \times 120 = 0.30 \times 120 = 36$. So 36 students own cats.

Well done checking your answers!

Keep practicing to strengthen your skills.

Author's Final Note

I hope you enjoyed this book as much as I enjoyed writing it. Whether you are a student working through the material, a parent supporting your child's learning, or a teacher guiding your class, I have tried to make this book as clear and engaging as possible. I hope I have succeeded. If you have any suggestions for improvement, please let me know. I would love to hear from you.

The accuracy of calculations is very important to me. We have done our best, but I also expect that I have made some minor errors. Constant improvement is the name of the game. If you find any errors, please let me know. I will fix them in the next edition.

For students: Your learning journey does not end here. I have written a series of books to help you learn math. Make sure you browse through them. I especially recommend workbooks and practice tests to help you prepare for your exams.

For parents: Thank you for investing in your child's education. I encourage you to explore the companion resources available online to help support your child outside the classroom.

For teachers: Thank you for the invaluable work you do every day. I hope this book serves as a useful resource in your classroom. Feel free to reach out if you have suggestions or would like to discuss how best to use this book with your students.

I also enjoy reading your reviews. If you have a moment, please leave a review on where you found this book. It will help others find this book. If you have any questions or comments, please feel free to contact me at drNazari@ViewMath.com.

And one last thing: Remember to use online resources for additional help. I recommend using the resources on https://ViewMath.com You can find video lessons, practice problems, and more. You can also use the online companion for this book to track your progress and access additional resources.

Wishing all students the best in their studies, parents every success in supporting their children, and teachers continued inspiration in their classrooms!

Dr. A. Nazari

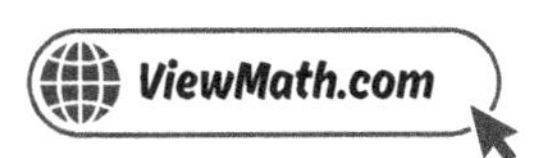

 # Great Job! Keep Learning with ViewMath!

Keep up the great work! Visit **viewmath.com/VA-Grade6** for free lessons, quizzes, and more.

Study Guide

Workbook

Step-by-Step

3 Practice Tests

5 Practice Tests

7 Practice Tests

Find more at
ViewMath.com/VA-Grade6

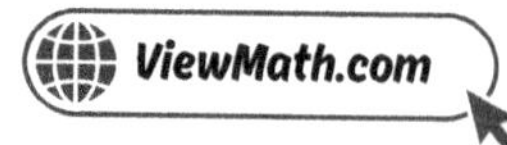